From the Airport to the City

Salk International Travel Premiums, Inc.

Houghton Mifflin Company
Boston / New York / London 1992

Copyright © 1991 by Salk International Travel Premiums, Inc.
All rights reserved

Published in previous editions as *Crampton's International Airport Transit Guide.*

For information about permission to reproduce selections from this book, write to Permissions, Houghton Mifflin Company, 215 Park Avenue South, New York, New York 10003.

ISBN 0-395-63321-4

Printed in the United States of America

INTRODUCTION

> "There is an air of absolute finality about the end of a flight through darkness. The whole scheme of things with which you have lived acutely, during hours of roaring sound in an element altogether detached from the world, ceases abruptly. The plane noses groundward, the wings strain to the firmer cushion of earthbound air, wheels touch, and the engine sighs into silence. The dream of flight is suddenly gone before the mundane realities of growing grass and swirling dust. ..."—Beryl Markham in *West With the Night.*

Ah, the dream of flight! How fast it fades when the wheels touch the runway. In the air, others worry about getting you safely from point A to point B. But on the ground you're on your own. And unless you're returning home or being met by a welcoming committee, getting from the airport to point C may require information you don't have. That's what this book is for.

At most airports there is more than one way to get to your final destination—taxi, airport shuttle, public bus, metro, helicopter, water taxi, rental car. The right choice will make the best match with your schedule and budget. This guide will help you make the right choice by providing the data you need for an informed decision: departure and running times, costs, frequencies, routes, and all the other details of airport ground transportation.

Symbols & Assumptions

This eleventh annual edition of *From the Airport to the City* contains data for 388 airports, arranged alphabetically by the name of the principal nearby city. The information is presented in the same sequence at each entry, beginning with the name of the airport and the distance and direction from the city.

Symbols—the same ones you will see in many airports—are used to identify the transportation options. Following is a key to the symbols including an explanation of the kinds of information presented at each, as well as certain assumptions that apply throughout.

Taxicab The first fare quoted is the cost to city center of the principal nearby city, in local currency with U.S. translation. Then: extra charges, if any; and tipping custom if there might be some question about what to give the driver. Assumptions: cab is metered; large enough for four, possibly five passengers with baggage; and one fare covers all. Exceptions are noted.

Airport Coach, Shuttle, Limo, Van Details: name of service, operating hours and schedule, route, fare, conveniences. Assume that this service, like the others, departs from or adjacent to baggage claim. Driver usually hoists baggage on and off as part of routine service. Hotels and transportation centers are the principal stops.

Public Bus Number, name, schedule, travel time, fare, baggage space if any. These buses usually make many stops on their way into the city; they seldom serve hotels. But in selected cities–London and Rome, for example–riding the public bus from the airport provides a close-up look at city life that might pass by too quickly in a cab, for example.

Metro, Subway, Tram, Train Destination, frequency, operating hours, travel time, fare, baggage space, comparative comfort. Numerous European cities have laid rail to the airport and shown what a good means of transportation this can be. Brussels, Geneva, London-Heathrow, Amsterdam, Frankfurt, and Barcelona are just a few examples. The United States has fewer airport rail links but some very good services. In Chicago, Cleveland, Philadelphia, and Atlanta, the rail station is within the airport. In Boston, Oakland, Washington, D.C., and New York-JFK, a good rail connection is just a short shuttle ride away.

Helicopter Schedule, fare. Such air taxis tend to come and go. This edition lists service at Boston, New York-JFK, and Los Angeles.

Rental Cars Companies with counters in the terminal as well as others serving the airport are listed, alphabetically.

P Parking Daily short-term (ST) and long-term (LT) rates are given in local and U.S. currency. Assume there is a free shuttle bus from baggage claim to the more distant, long-term lot.

Free Transport

One very common means of airport ground transportation is not listed here. It is the free shuttle service provided by many hotels and motels. So if you're headed for a hostelry, check the bank of "Courtesy" telephones in baggage claim.

Accuracy

We have taken care to present accurate information. But remember that services, schedules, travel times, fares, and foreign exchange rates change from time to time. If certain data are critical to your travel plans, reconfirm.

In certain countries where inflation causes frequent changes in local prices or where the U.S. dollar is readily accepted, fares are quoted in U.S. dollars only.

Feedback

We rely on feedback from travelers to keep the listings current. If your experience differs from what is indicated here, or if you discover new options for getting away from the airport, please let us know.

The Editors

SALK INTERNATIONAL TRAVEL PREMIUMS, INC.
P.O. Box 1388
Sunset Beach, CA 90742

Telephone: (714) 893-0812
FAX: (714) 556-7255

ACKNOWLEDGMENTS

Many people have contributed to the completeness and accuracy of the listings. The editors wish to recognize the valuable assistance provided by the following:

United States, Canada Gordon Allison, C.M. Amour, Wayne Anderson, Sandra Appler, Dana Banks, John Barsalou, David Blackshear, Daniel Booz, Lindy Boues, Larry Bowers, Janis Brand, Hoyt Brown, Robert Brown, Sara Brownlowe, Robert Bullock, Ed Burke, Larry Burke, Ann Campo, Sharon Caudill, Myron Clement, Dee Dee Corey, Shawn Dobberstein, Lynne Douglas, Nancy Duncalfe, Louis Fair, Carolyn Fennell, Robert Flannery, Pam Freeman, Barry Fukunaga, John Garwin, Laura Gipson, Steven Goldman, Michelle Grage, Peter Greenstein, Daniel Grining, John Halloran, John Hanks, Janet Hansen, Joyce Harvard, Darlene Hill, Kevin Hill, Joseph Hills, Donald Hobbs, Donald Hoeft, Christina Holden, Joaquine Jones, Karon Keith, Amyn Keshavjee, Lawrence Krauter, Marilyn Lands, Tracy Lane, Sharon Ledward, R. Maguire, Mary Lou Mania, Tracy Marshall, Richard McCollum, Shana McDonald, Floyd McKenzie, Rod McLean, Keith Meuclin, R. Michael, Nancy Milton, Elly Mixsell, Rod Paine, Jim Parker, Laurie Patterson, Adrienne Pittman, S. Ransom, B. Rapier, Gerald Reas, Loren Riley, Alesia Robinson, Sam Samaddar, O.J. (Jim) Schaller, Don Schult, Kevin Shirer, Jacqueline Shuck, Jenny Stacy, Christine Stern, Shirley Tessmer, Patricia Testa, Karen Thomas, Louise Thomassin, Warren Thompson, Luther Trent, Billy Tucker, Betsy Wade, Terry Wagner, Jeannie Weiss, Courtney Wiercioch, James Wilson, Janet Woods, Nordean Yates, Janice Young, Rochelle Young.

Caribbean, Central & South America Jean Bozzuto, Glenn Holm, Barry Hutchinson, Motiarn Janiel, Willian Jedwab, Elton Jones, Michael Nicolaas, Oswaldo Sansone, Kelly Walsh.

Europe, Africa Abdullah Abdullkafe, Norman Allan, Dr. Amore, Norman Bartlett, C. Bornand, Fatima Boul-Ela, Hilga Brenner-Khan, Michael Briers, Orhan Burcak, Antonio Carualho, Ildefonsa Costa, Arthur Coughian, Julie Cullen, Georges Daverat, Fiona Donomer, George Dyce, B. Frivold,

J.P. Garnier, Verona Giuseppe, Okan Gokboru, Max Haberstroh, Ellen Hacker, Peter Hanakovic, Bo Haugaard, Annemiek Holkeboer, Patricia Houzelle, Vivienne Hunnisett, M. Kalita, Dr. La Ganga, Peter Leonard, Simon Leysmon, Luis Hernandez Lorente, Joan Marso, Robert Payne, M. Pukljak, Olafur Ragnars, Michael Tmej, Jackie Tyrrell, Dominic Vaughan, Elisabet Wahrby, Simon Wamai, Sandra Watt, Wim Wegter.

Far East, Australia, New Zealand, Pacific Phillip Cash, A. Clement, Lyn Greenfield, Teng Hong Hai, Haji Jamaludin, Chang Hyo Kang, Angela Lark, Qian Nailin, Mandy Needle, Gp. Capt. Charoon Peetong, Carol Wang, Junko Watanabe, Wychiffe Wewa, Roger Yang.

ABERDEEN, Scotland
Aberdeen/Dyce Airport, 7 mi (11 km) NW

£6 ($10), 15-25 min. Tip 10%.

No. 27 every 30-40 min 6:40 am-10:15 pm M-F, less frequently on weekends. 34 min to Guild St. Aberdeen, next to RR & bus stations. 65p ($1).

Alamo, Avis, Budget, Eurodollar, Europcar, Hertz, Kenning.

P £4.00 ($7)/day first six days, £2.50 ($4)/day each day thereafter.

ABIDJAN, Ivory Coast
Aeroport Intl d'Abidjan-Port Bouet, 9.3 mi (15 km) SE

Hotel Ivoire $7, 30 min; Hilton and Novotel $5, 20 min. Fares are doubled between midnight and early morning.

Avis, Budget, Europcar, Hertz.

ABU DHABI, United Arab Emirates
New Intl Airport, 12.5 mi (20 km) S

Di20-25 ($5.40-6.75), 20 min to city center by Al ghazal taxi. Confirm fare with driver before you get in the cab. Tip 10%. Return fares from hotels slightly higher.

Avis, Budget, Europcar, InterRent, National, Sunny.

ACAPULCO, Mexico
Juan N. Alvarez Airport, 12 mi (20 km) SE

$8, 30 min to La Condesa Beach area hotels.

Servicio Colectivo about every 10 min from 7 am to last flight. Stops anyplace on request along route through city. $3. Look for vehicle in front of terminal.

Avis, Dollar, Hertz, Sands.

ADELAIDE, Australia
Adelaide Airport, 5 mi (8 km) W

A$7 ($5.40), 10-15 min.

Transit Coach Service from curb outside terminals every 30 min 6 am-9 pm to Hilton Intl, Richmond, other major hotels. A$1.50 ($1.15), 10-15 min.

Action, Avis, Budget, Hertz, Koala, Thrifty.

P A$6 ($4.60).

AKRON-CANTON, Ohio

Akron-Canton Regional Airport, 15 mi SE of Akron, 10 mi NW of Canton

$14 flat rate, 15-20 min to Canton; $32 flat rate, 20-30 min to Akron. (If no cab at airport call Canton Yellow, 216-456-4343. May take 10 or more min to arrive.)

Hopkins Limousine to Quaker Square, Holiday Cascade downtown. Continues to Holiday Inn, Hilton in Fairlawn. $8.25. M-F at 5:15, 6:25, 7:35, 9:15, 10:30, 11:45 am; 1, 2:15, 3:30, 4:45, 6, 7:15 pm. Sun from 10:30 am. Sat-Hol 6:05 am, 12 noon only. Service also to Ramada Inn Northwest, Best Western Medina, Cleveland-Hopkins Airport. Info: 800-543-9912 or 216-362-3795.

Avis, Budget, Dollar, Hertz, National, Snappy, Thrifty.

P ST $5, LT $3.75.

ALBANY, Georgia

Southwest Georgia Regional Airport, 4 mi SW

$4, 5-10 min.

Avis, Budget, Hertz, National.

P $4/day.

ALBANY, New York

Albany County Airport, 10 mi NW

$13 flat rate one person; more than one psgr, $8.50 each. 18-20 min to Capitol.

Airport Limousine operates 6 am-12 am. To Capitol area, $13 one psgr, $8.50 per psgr two or more. Info: 518-869-2258.

No. 1 Central bus every 30-60 min 5:43 am-6:07 pm. 60¢. 45-50 min run to Broadway, downtown.

Ajax, American Intl, Avis, Budget, Dollar, Hertz, National, Snappy, Thrifty.

P ST $9.75, LT $5.50.

ALBUQUERQUE, New Mexico
Albuquerque Intl Airport, 5 mi SE

$8-10, Addl riders 50¢ each, 10-20 min.

Yellow Cab van, same fare as cab.

No. 50 bus every 30 min 6:37 am-6:07 pm M-F. Sat hourly 6 am-5 pm. No Sun, Hol. 30 min to Fifth & Gold, downtown, 75¢.

Alamo, American Intl, Avis, Budget, Dollar, Enterprise, General, Hertz, National, Snappy, Thrifty.

P ST $5, LT $4.

To Santa Fe See Shuttlejack listing at Santa Fe. **To Las Vegas, NM** TNM&O Airport Express at 11:30 am. $21. Info: 505-758-1144.

ALEXANDRIA, Louisiana
Esler Regional Airport, 15 mi NE

$20 flat rate downtown, 25 min.

Airport shuttle meets flights, $7.50-10.50, 20 min.

Avis, Budget, National.

ALICE SPRINGS, Australia
Alice Springs Airport, 9 mi (15 km) S

A$17-18 ($13.10-$13.90), 15 min.

Alice Springs Airport & Railway Shuttle Service meets all arriving flights; A$4 ($3.10), 15 min.

Avis, Brits, Budget, Hertz, Territory.

P A $3/day.

ALLENTOWN, Pennsylvania
Allentown-Bethlehem-Easton Airport, 5 mi (8 km) NE

$10-12, 15-20 min. To Bethlehem, $12-15, 20 min; to Easton, $25, 30 min.

Northampton Transit Authority (LANTA), 6 am-6 pm, $1, 25 min.

Avis, Budget, Hertz, National, Snappy, Thrifty.

AMARILLO, Texas
Amarillo Intl Airport, 9 mi E

$12-13, extra riders 50¢ each. 15-20 min.

Avis, Budget, Hertz, National.

AMSTERDAM, Netherlands
Amsterdam Airport Schiphol, 9 mi (15 km) SW

Dfl47-52 ($25-28), 20-25 min. To The Hague, Dfl115 ($61), 35 min; Rotterdam, Dfl150 ($80), 55 min; Utrecht, Dfl115 ($61). Cabs seat 4-6 persons. Meter fare includes tip and luggage but round up to next guilder for short ride, add 10% for longer hauls.

KLM Yellowline bus to nine major hotels every 30 min 6:30 am-9:45 pm, then 10:20, 11, 11:40 pm. Hilton Amsterdam, Barbizon, Park Apollo, others. Dfl15 ($7.80). Free to Ibis. 25-35 min. KLM Orangeline bus to five hotels every 30 min 6 am-10 pm, then 10:40, 11:15 pm. Serves Pulitzer, Krasnapolski, Holiday Inn, Victoria, Barbizon Palace, Amsterdam Sonesta, Golden Tulip. Dfl15 ($7.80). Purchase ticket from driver or road transport desk in airport. Good baggage space on both buses, comfortable.

To Amsterdam Central Station, every 15 min 5 am-12:05 am, then hourly. Dfl4.50 ($2.35), 17-20 min. Modern, comfortable cars, plenty of baggage space. Railway station is in basement, one level below arrivals area. Buy ticket before boarding. Schiphol is on the Dutch intercity rail network. Besides direct service to center city, World Trade, and RAI Convention Centre, express service is available to Rotterdam, Den Haag, Groningen, Leeuwarden, Vlissingen, and other points throughout Western Europe.

No. 173 to RAI, Amstel stations every 15-30 min 6:24 am-6:24 pm, 7:10 pm then hourly to 12:10 am. 60-min ride.

Ansa, Avis, Budget, Eurodollar, Europcar, Hertz, InterRent, Van Wijk.

P ST Dfl25 ($13), LT Dfl5 ($3).

Schiphol sets the world standard for airport signage, both inside and outside the terminal.

ANCHORAGE, Alaska
Anchorage Intl Airport, 7 mi (11 km) SW

$12-13, 10-22 min.

Dynair Limousine van hourly 5 am-2 am. $6. To Captain Cook, Anchorage Westward Hilton, Holiday Inn, Sheraton, other hotels. Also serves Palmer, Wasilla, Eagle River.

No. 6 People Mover from lower level hourly 7:20 am-7:50 pm M-F. 75¢. 20-25 min downtown.

Alamo, Avis, Budget, Hertz, National, Payless.

P ST $7, LT $5, International $5/day, $25/week.

ANGUILLA, Leeward Islands, West Indies

Wallblake Airport, 2 mi (3.2 km) SW of the Valley

To hotels: Malliouhana, 6¼ mi, $12 (US currency accepted); Cinnamon Reef, 2½ mi, $7; Mariners, 3½ mi, $8; Cove Castles, 8½ mi, $15; Shoal Bay Villas, 4½ mi, $8; Anguilla Great House, 8½ mi, $10; Coccoloba, 9 mi, $15; Carimar Beach Resort, 7 mi $12; Rendezvous Bay Hotel, 8½ mi, $10; Cap Juluca, 8 mi, $15.

Avis, Budget, Connors, H&R.

ANKARA, Turkey

Esenboga Intl Airport, 18 mi (29 km) N

Tl80,000 ($17.50) flat rate for up to 3 persons. 30-min.

Airport Bus operated by Turkish Airlines meets flights. Tl12,000 ($2.60). 30 min.

Avis, Budget, Camel, Hertz.

APPLETON, Wisconsin

Outagamie County Airport, 4 mi W

$8 flat rate one person, $1 each addl. 12 min.

Avis, Budget, Hertz, National.

ASHEVILLE, North Carolina

Asheville Regional Airport, 12 mi S

$15 flat rate, 15-20 min.

To Asheville, $12, 15-20 min; to Hendersonville, $12, 15 min; to Black Mtn/Brevard, $35, 20 min; to Waynesville, $65, 35 min.

Avis, Hertz, National.

P ST $4.50, LT $3.75.

ASPEN, Colorado

Aspen/Pitkin County Airport, 5 mi W

$12, 10 min; Snowmass, $25, 15 min; Highlands, $8-9, 10 min; Starwood, $10-12, 10 min. Addl passengers 50¢ ea; baggage 50¢ ea; airport departure fee 50¢.

To Aspen, $5/person (3 person minimum); Snowmass $6.50/person.

Roaring Fork Transit Agency bus departs hourly 7 am-1 am during ski season from highway in front of terminal. $2 (50¢ from hwy), 5-10 min. Baggage OK.

Alamo, Avis, Budget, Dollar, Eagle, Hertz, Mountain Express, National, Thrifty.

P ST $3.50, weekly $17.50.

ATHENS, Greece

Hellinikon Airport, 6.5 mi (11 km) S

Dr1000-1500 ($5.30-8), 30 min downtown. Tip 10%. Dr50 (25¢) each bag. Double fare 1-5 am. Between terminals, a 7-min taxi ride. Also a bus every 20 min 6 am-midnight, Dr160 (85¢); midnight-6 am, Dr200 ($1.05). Cabs not air conditioned. At start of ride, make sure taxi meter set to "tariff 1" (or "tariff 2" during double fare hours, 1-5 am).

West Air Terminal, Olympic Airways

Olympic Airways bus every 30 min to city terminal on Syngrou Ave. Dr160 (85¢).

No. 133 from 6:10 am-12:30 am. Dr50 (30¢). 20-25 min to Syntagma (Constitution) Square. To Piraeus: No. 109 every 20 min 6 am-midnight. Dr50 (30¢).

East Air Terminal

To Vassilisis Amalias Ave. every 20 min 6 am-midnight. Dr160 (85¢). To Piraeus: No. 101 every 20 min 5 am-10:45 pm. Dr50 (30¢).

Avis, Budget, Eurodollar, Hellas, Hertz, InterRent/Batek, and others.

P Dr140 (75¢).

ATLANTA, Georgia

Atlanta Intl Airport, 8 mi SW

$15 flat rate one passenger, $8 each for two, $6 each for three. 20-30 min. **To Buckhead business area, Lenox** $25 one passenger, $13 each for two, $9 each for three or more.

Atlanta Airport Shuttle every 20 min 5 am-midnight from ground transportation area. $7 OW $12 RT, 35-40 min to downtown hotels. Hourly service to Lenox, Buckhead, Emory areas. $10 OW, $18 RT. Info: 404-766-5312. **To northern, eastern suburbs** Northside Airport Express. **Dunwoody** Radisson, Hyatt Ravinia, Guest Quarters, Marriott Suites, Marriott Perimeter. Every 45-60 min 6:45 am-11:15 pm. $10.25 OW, $18.75 RT. **Windy Hill** Marriott Northwest. $9.95 OW, $17.95 RT. Every 45-60 min 6:45 am-11:15 pm. **Roswell** Every 60 min 6:45 am-11:30 pm. $15 OW, $25 RT. **Gwinnett** Gwinnett Marriott. $13.75 OW, $25 RT. Every 60 min from 7 am; last at 10:45 pm. **Memorial** Every 60 min from 7 am; last at 10:45 pm. $9.95 OW, $17.95 RT. Info: 404-455-1600.

MARTA train from station next to baggage claim, South Terminal. Fare 85¢. Clean, quiet, air conditioned cars. Space to stow luggage and hang garment bags. 15-min ride to Five Points, downtown. Next stop, Peachtree Center, is a block or two walk from Ritz Carlton, Westin, Hyatt Regency, Barclay, Atlantan, Marriott, Best Western. Civic Center is the following stop, 17 min from airport. Trains leave every 6-12 min M-F, 4:35 am-1:17 am; every 10-15 min Sat, 4:44 am-1:06 am; every 15 min Sun, 5:17 am-12:47 am.

Alamo, Avis, Budget, Dollar, General, Hertz, Major, National, Value.

P ST $12, LT $3-5.

To Roswell Roswell Airport Express hourly 6:45 am-11:30 pm to Crabapple Square. $15 OW, $25 RT. Info: 404-998-1893. **To Macon, Warner Robins** Groome Trans. hourly 9 am-9 pm. 90-min run to Shoney's Inn; then to Hilton, Warner Robins. Service every 2 hrs Sun am, all day Sat. $17.50. **To Marietta, Peachtree Corners** Airport Connection hourly 7:45 am-11:30 pm. To Sheraton Parkway 75 Hotel, Marietta, $13 OW, $22 RT. To Courtyard Marriott, Hilton, Marriott at Peachtree Corners, $18 OW, $30 RT. Info: 404-448-1544. **To Rockdale, Newton counties** East Metro Airport Shuttle. Res & info: 404-929-8044.

ATLANTIC CITY, New Jersey

Bader Field, 1 mi W

$4-10, 5-10 min to casinos.

Avis, Budget, Hertz, Snappy.

ATLANTIC CITY, New Jersey
Atlantic City Intl Airport (Pomona Field), 13 mi NW

$30-35, 20 min.

Sterling Shuttle, at Gate 3, to all hotels and casinos. $7.50, 25 min.

Avis, Budget, Hertz.

AUCKLAND, New Zealand
Auckland Intl Airport, 14 mi (21 km) S

NZ$32 ($18.50), 35 min. 10 pm-6 am and all day Sat, Sun, Hol: NZ$35 ($20.20). No tip.

Airporter from Intl terminal every 30 min 6:30 am-8:30 pm. Stops at domestic terminal, Travelodge, any bus stop en route, Sheraton-Auckland, Hyatt Kingsgate, Farthings hotels (opposite RR station). 35-40 min to last stop, downtown terminal. NZ$9 ($5.20).

Shuttles operates 5 am-midnight (or last plane). NZ$12 ($6.90) one passenger, NZ$7 ($4) each, two or more.

Avis, Budget, Hertz, Thrifty.

P NZ$11 ($6).

AUSTIN, Texas
Robt. Mueller Municipal Airport, 4 mi NE of State Capitol

$7, 8-9 min.

No. 20 bus at sign on center median outside baggage claim. Every 15-30 min 5:29 am-11:36 pm. 50¢. 22 min to Capitol, downtown.

Advantage, Alamo, Avis, Budget, Dollar, General, Hertz, National, Thrifty.

P ST $9, LT $5.50/$4.75/$2.75.

AYERS ROCK, Australia
Connellan Airport, 4 mi (6 km) N Yulara, 16 mi (26 km) NW Ayers Rock

Deluxe Coachlines meets flights; A$7 ($5.60), 10 min to Sheraton, Four Seasons, Campground, Lodge.

Avis, Budget, Hertz, Rock 'N' Ride, Thrifty.

BAHRAIN ISLAND, Bahrain

Bahrain Intl Airport, Muharraq, 5 mi (8 km) NE

Sign posted at taxi ranks shows average fares to various destinations but expect to pay a little more. Across causeway to Manamah, D5 ($13.25), surcharge after midnight. No meters. Authorized cabs bear orange stripe; avoid others.

Arabous Car Hiring, Avis, Budget, Europcar, Manzuri Transportation Service, Marsha Car Hiring.

BAKERSFIELD, California

Meadows Field-Kern County Airport, 4 mi NW

$9, addl riders free. 10-15 min.

Route 3 hourly 7:15 am-6:15 pm. 50¢.

Avis, Budget, Hertz, National, Thrifty.

BALI (Denpasar), Indonesia

Ngurah Rai Airport, 8 mi (13 km) SW

Cabs are blue. To Denpasar, Rp9000 ($5), 25 min; to Kuta Beach, Rp4500 ($2.50), 10 min; to Sanur Beach, Rp12,000 ($6.50), 25 min; to Nusa Dua Beach, Rp12,000 ($6.50), 20 min. Three-passenger limit to cab. Tip optional.

Some hotels provide free minibuses. Inquire.

Bali Beringin, Bali Giri, Bali Happy, Bali Wisata at Kuta Beach. International license required.

P 12¢/day.

BALTIMORE, Maryland

Baltimore-Washington Intl Airport, 10 mi S

$16, 15-25 min to downtown Baltimore. Annapolis $34; Washington $42, 45 min.

BWI Airport bus/van, $6.50 OW/$12 RT. Every 30 min 7 am-11 pm. To Stouffer, Hyatt, Harbor, Sheraton, Omni Lord Baltimore, Days Inn, Holiday Inn, Marriott. **To Annapolis** Hourly 7 am-11 pm. $12 OW, $22 RT. Buy ticket at Ground Transportation Center.

No. 230 MTA bus at sign, lower level. Operates M-F at 6:08, 6:49, 7:24, 8:04, 9:04 & every 40 min. Last at 3:44. $1.85 exact change. About 30 min to Howard & Lombard Sts. downtown.

To National Airport, Washington Bus every 90 min 7 am-10 pm weekdays, Sat & holidays, 7:30 am-8 pm, $13. This service takes you to 16th & K St. terminal in downtown D.C.. From there you take cab to National, about $8 addl fare. Info: 301-859-7545.

Alamo, Avis, Budget, Dollar, Hertz, National, Thrifty.

P ST $13, LT $8/$5. Valet parking $15 first day, $9 each addl.

To Amtrak Free shuttle from airport to rail terminal 1.5 mi distant. Connections to Baltimore, Washington, New York. Info: 800-USA-RAIL and at Ground Transportation Counter, Central Terminal.

BANGKOK, Thailand

Bangkok Intl Airport, 13 mi (22 km) N

Thai Intl Airways operates cabs marked by their logo. Flat rate B300 ($11.60), no extras. No tip. A/C. Buy ticket at desk outside customs. 40-min trip. Regular cabs do not have meters: take time to agree on fare before getting in. To downtown hotels, fares should range B250-300 ($9.70-11.60).

Airport shuttle every 30 min 7 am-8 pm. Fare B60 ($2). Buy ticket at desk outside customs hall. Makes hotel stops. (Asia and Viengtai Hotel).

No. 4, 10, or 13 are air conditioned, fare B5-15 (20-60¢). No. 29, 59 are not a/c, fare B2-3 (8¢). Frequent service. Bargain trip to city center but jam packed during rush hours.

Avis, Hertz, Toyota. (City offices only.)

BANGOR, Maine

Bangor Intl Airport, 3 mi W

$5, addl riders 25¢ each, 8-10 min.

City Bus every 30 min 6:18 am-5:18 pm M-Sat. 60¢, 15 min. Flag bus down as it passes terminal.

Avis, Budget, Dollar, Hertz, National, Thrifty.

P $4.50/day, $22.50/week.

BARCELONA, Spain

Barcelona Airport, 6 mi (10 km) SW

Pta1500 ($14.25), 15-20 min. Tip 10%.

RENFE train every 30 min 6 am-11 pm. Pay fare Pta130 ($1.25) before boarding. Though station is half a mile from terminal, moving sidewalks make the connection easy. Follow pictograms. Roomy seats, plenty of baggage space. 16 min to Sants (Central) Station.

EA/EN bus approx hourly 6:35 am-2:30 am. Stops at yellow sign at either end of terminal. Fare Pta125 ($1.20) paid on board. 28 min to Plaza Espana, Metro connection. No baggage space, slow; but good way to see postwar sections of Barcelona.

Atesa, Avis, Europcar, Hertz.

P Pta405 ($3.85)/day.

BASEL, Switzerland

Basel/Mulhouse Airport, 6 mi (10 km) NW of Basel

SF30 ($20) to Hotel Schweizerhof, 15 min. To Mulhouse FF250 ($45), 20-30 min.

Swissair bus to Basel main railroad station. Pay fare SF5 ($3.40) Bus meets all flights. 15-20 min ride.

Avis, Budget, Europcar, Hertz, InterRent, Milleville.

P French side, F34 ($6). Swiss side, SF15 ($10).

BATON ROUGE, Louisiana

Baton Rouge Metropolitan (Ryan Field), 8 mi NE

$16, 15-20 min.

Avis, Budget, Hertz, National, Snappy, Thrifty.

P ST $6, LT $4.50.

BEIJING, China

Beijing Capital Airport, 18 mi (29 km) NE

RY42 ($8), 35-40 min. To Great Wall Hotel RY35 ($6.70), 30 min; Li Do Holiday Inn RY14 ($3), 20 min. No tip. English speaking taxi company is Capital Taxi, telephone 55-74-61. Be prepared to tell where you are located, where you are going.

CAAC (Civil Aviation Administration of China) bus to city ticket office RY8 ($1.50). Meets flights. 40 min. It may be more convenient to take cab direct from airport rather than try to find one in city after leaving CAAC bus.

No. 359 bus from airport to Dongzhimen, center city. Every 10 min 5:10 am-9:30 pm, (15¢) 45 min.

Self-drive rentals not available in China. National Car Rental provides cars with drivers. Office: South Xin Hua Street, He Ping Men, Beijing. Tel: 33-13-49. Also, taxis always available in front of big hotels.

BELFAST, Northern Ireland
Belfast Intl Airport, Aldergrove, 18 mi (29 km) NW

Approx £16 ($25). £2 ($3.15) surcharge 11 pm-6:30 am. 20 min.

Ulsterbus Express to main bus stations in Oxford and Great Victoria Sts. Every 30 min from 6:45 am M-Sat. Last, 10:10 pm. 7:15 am-10:15 pm Sun. Baggage space. Fare £3 ($4.70). 35-40 min trip. Follow "Way Out" signs from baggage claim. Bus sits facing exit doorway.

Avis, Godfrey Davis/Europcar, Hertz, McCausland.

P ST £20 ($31.60), LT £2.50 ($3.95).

BELGRADE, Yugoslavia
Airport Beograd, 10 mi (16 km) W

$6 to center city, RR station, JAT terminal. Baggage 15¢ ea. Tip 10%. 20-min ride.

Service every 15-20 min to Hotel Slavija, RR station, JAT terminal. $1.50. 35 min.

Alptour, AMSJ, Avis-Autotehna, Globtour, Hertz-Kompas, Inex, InterRent, Putnik, Unis.

P $4 day.

BERGEN, Norway
Bergen Airport, 11 mi (19km) SW

NKr170 ($25.50), 20 min. Tip 10%.

Meets arriving SAS and Braathens SAFE flights. NKr36 ($5.40); 30 min to bus terminal, then to Norge Hotel, SAS Royal Hotel. Return service every 30 min 5:50 am-9:50 pm. Outbound service starts at hotels, then picks up at bus terminal, where coaches may become crowded. Baggage racks.

Avis, Budget, Hertz, InterRent/Europcar, Thrifty.

P NKr45 ($3.80)/day.

BERLIN, Germany
Schonefeld Airport, 11 mi (18 km) SE

30 min. Arrange for taxi hire at Verkehrsbetriebe (airport transit service), first floor of terminal.

Interflug operates a shuttle bus every half hour, terminating at Berlin Funkturm. 60-min ride.

Trolleys every 20 min, more frequently in peak periods, stopping at Alexanderplatz, Marx Engels Platz, and terminating at Friedrichstrasse, Berlin. 45-min ride to town. Station is 500 meters from terminal—shuttle bus available.

Hertz, Rent-A-Car.

BERLIN, Germany
Tegel Airport, 4 mi (7 km) NW

DM17-19 ($10-11). Tip 10%. DM1 (60¢) per bag.

No. 9 Airport City Bus from Gate 8, arrivals area. Also stops at upper level Main Hall. Leaves every 7½ to 15 min 4:50 am-midnight. Fare DM2.70 ($1.60). Pay driver or use ticket-vending machine inside airport. 24-hr ticket can be purchased at airport or in city for DM9 ($5.20). Comfortable seating, baggage racks. Route into city is Kurfurstendamm. 30 min to Zoo—central train station for West Berlin. 8 min after leaving airport, bus stops at U-Bahn Jacob Kaiser Platz where bus ticket may be used as free transfer to U-Bahn, S-Bahn system.

Autohansa, Avis, Europcar, Hertz, InterRent, Sixt-Budget, Westfehling.

P ST DM13 ($7.50), LT DM10 ($5.80).

BILLINGS, Montana
Billings Logan Intl Airport, 2 mi NW

$3 flat rate one person; $2.25 each addl. 10 min.

Avis, Budget, Dollar, Hertz, National, Thrifty.

BIRMINGHAM, Alabama
Birmingham Municipal Airport, 5 mi NE

$8-$9 one person, $5 each flat rate two or more, 15 min.

No. 20 Zion City bus. Every 30-45 min M-Sat, 6 am-6 pm. 80¢. 20 min run to 3rd Ave. & 21st St.

Avis, Budget, Dollar, Hertz, National, Payless, Snappy, Thrifty.

P ST $6.50, LT $3.

BIRMINGHAM, England
Birmingham Intl Airport, 7 mi (11 km) SE

£8-10 ($13.40-16.80) downtown, 20-35 min. Tip 15%. There is a surcharge for trips terminating more than 16 km from Birmingham center.

Air terminal is less than 1 km from British Rail's Birmingham Intl station. Good connections into Birmingham, other Midlands cities. Follow "Transit Link" signs first floor. Get aboard MAGLEV —departures every 5 min. 2-min ride to train station. Service to central Birmingham every 20 min 6 am-10 pm, then every 35 min until midnight. Sunday service about every 45 min. Fare £1.35 ($2.30). 15 min to Birmingham New Street, downtown. "Travel Centre" in rail station will help with other BritRail connections.

No. 900 West Midlands Travel from shelter at second curb outside terminal. Service Mon-Sat every 20 min 6 am-7:49 pm. Then every 30 min to 11:14 pm, last 11:49 pm. Sun hourly 6 am-10:14 pm. Then every 30 min to 11:14 pm, last at 11:49 pm. Fare 95p ($1.65).

Avis, Europcar, Hertz.

P ST £11 ($19.15), LT £2.50 ($4.35).

BISMARCK, North Dakota
Bismarck Municipal Airport, 3 mi SE

$5, 10 min; extra passengers, $1 ea.

Avis, Budget, Hertz, National.

P $3.75/day, $18.50/week.

BOGOTA, Colombia
El Dorado Intl Airport, 8.5 mi (14 km) E

$2.50, 30-40 min to city center. Police dispatcher gives passengers a ticket with fare to destination. This amount is paid to driver. Beware unlicensed cabs operating late at night after dispatch system closes.

Avis, Budget, Hertz, National.

BOISE, Idaho
Boise Air Terminal, 3.5 mi S

$8, $1 extra for more than two riders. 10-15 min.

Avis, Budget, Dollar, Hertz, National, Payless, Thrifty.

BOMBAY, India

Bombay Intl Airport, 19 mi (30 km) N

R105 ($6), 50-75 min. To Juhu Beach, R46 ($2.60), 25 min; to Dadar, R60 ($3.50), 35 min. Small bag, R3 (20¢), large bag, R5 (30¢). Four passengers maximum. Tip not expected.

Ex-Servicemen's Air Link Transport Service (EATS) on the hour from intl, domestic terminals except midnight, 1 am, 6 am. R25 ($1.40). 60 min ride to Air India Building, Nariman Point.

Not available in India.

P R8 (50¢)/day.

BONN, Germany See Cologne/Bonn, Germany

BORDEAUX, France

Bordeaux (Merignac) Intl Airport, 8 mi (13 km) W

F150 ($27), 20 min. Baggage F3 each. Tip 10%.

Navette Aeroport coach every 30 min 6:10 am-11:30 pm. Last departure Sat 11:10 pm; last Sun 11:10 pm. F30 ($5.40).

To Merignac CGFTE Bus No. 73 hourly 6:15 am-7:45 pm, Sun 7:50 am-7:40 pm. F14 ($2.50). Connects at Hotel des Postes with Service M to Bordeaux.

Avis, Budget-Milleville, Citer, Eurodollar, Europcar, Hertz.

P ST F57 ($10.25), LT F30 ($5.40).

BOSTON, Massachusetts

Logan Intl Airport, 3 mi (5 km) NE

$10 to downtown Boston, Cambridge. 15-25 min.

Downtown Airways red-white van hourly on the hour 8 am-10 pm to downtown hotels, $6.50-7.50. Up to hour travel time depending on traffic. City Transportation to downtown, every 30 min 7:15 am-9:15 pm.

Framingham-Logan Express Bus leaves Logan every 30 min M-F 6:30 am-11 pm & 11:45 pm. Sat 7 am-11 pm on the hour. Sun 7 am-1 pm on the hour, 1:30 pm-11 pm every 30 min & 11:45 pm. 45-60 min to parking lot at Rte 30, Shoppers World, Framingham. Early Bird Special, Sat only, 5:30 am leaving Framingham. Bus fare $8 (M-F), $7 (Sat-Sun) OW. Parking $4/day.

Braintree-Logan Express Pick-up lower level terminals A, B, C, E at Logan Express Bus Stop. M-F 6:30 am-8:00 pm every 30 min 8:00 pm-11:00 pm hourly on the hour; final bus 11:45 pm. Sat 7:00 am-11:00 pm; Sun 7 am-1 pm hourly, 1:00 pm-8 pm every 30 min, 8:00 pm-11:00 pm hourly; final bus 11:45 pm. 30-45 min, $5. Behind Sheraton Tara off Route 37 on Forbes Road across from South Shore Plaza. Parking $4/day.

Free bus No. 22 or 33 every 8-12 min to Airport Station of Rapid Transit (T). Clean, comfortable; baggage racks. At rail station, Blue Line train every 8-12 min 5:25 am-1 am. 85¢. Clean coaches, good wall maps. Four stops (20 min) to Government Center.

Airport Water Shuttle Free van every 5-10 min to Logan boat dock for water shuttle to Rowe's Wharf, next to Boston Harbor Hotel. Boat leaves every 15 min M-F 6 am-8 pm; every 30 min 12 noon-8 pm Sun. $7. 7 min across harbor. Fine way to start the business day in Boston!

Alamo, American Intl, Avis, Budget, Dollar, Hertz, National, Thrifty.

P $13 first day, $11 each add'l. day, $50/week.

To South Shore/Cape Cod Plymouth & Brockton Bus, 800-328-9997. **SE Mass, S. Suburbs** Bonanza Bus, 800-556-3815. Hudson Limo, 800-367-3885. **To Central Mass, Western Suburbs** Logan Express, 800-23-LOGAN. Peter Pan Bus, 426-7838. Hudson Limo. **To N Central Mass, NW Suburbs** Hudson Limo. McCarty Limo, 800-233-0066. Vermont Transit Bus, 800-451-3292. **To N Mass, N Suburbs** Vermont Transit Bus, McCarty Limo. Hudson Limo. **To N. Shore** C&J Trailways, 800-258-7111. Hudson Limo. **To New Hampshire** C&J Trailways. Concord Trailways Bus, 800-258-3722 (in Mass). Hudson Limo.

Toll-free U.S. number for Logan Airport info: 800-23-LOGAN At the airport, look for information booths opposite baggage claim in terminals A, C, E. Also interactive video display terminals nearby provide info 24 hours a day.

BOZEMAN, Montana

Gallatin Field, 8 mi NW

Cabs do not wait at airport. Call 586-2341. $9 flat rate, 15 min. Extra passengers, $3 ea.

Avis, Budget, Hertz, National.

P $4/day.

BRADENTON, Florida See SARASOTA/BRADENTON, Florida

BRATISLAVA, Czechoslovakia
Bratislava Ivanka Airport, 6 mi (10 km) NE

K120 ($5), 20 min. Tip 10%.

CSA-Bus meets arrivals, departures 6 am-midnight. Purchase ticket from driver, K5 (20¢). 30 min to city terminal near Hotel Palace, Devin.

No. 24 every 20 min 5 am-midnight from pick-up outside terminal. Pay driver K1 (5¢). 35 min to city. Baggage OK.

Avis, Europcar, Hertz, Pragocar.

BREMEN, Germany
Bremen/Neuenland Airport, 2 mi (3.5 km) S; DM12-15 ($7-8.70), 10 min.

No. 5 tram every 7-20 min 5:02 am-11:49 pm M-F. Sat-Sun service at 15-30 min intervals. DM2.4 ($1.40). To Bremen Hauptbahnhof in 15-20 min.

Avis, Europcar, Hertz, InterRent, Sixt Budget.

P DM6 ($3.50).

BRIDGEPORT, Connecticut
Sikorsky Memorial Airport, 10 mi S

$10.50 flat rate, 12-15 min. To Fairfield $12-18, 20 min; Milford $12-22, 12-20 min; New Haven $28-41.50, 20-25 min.

Airport Taxi, $10.50. Operates 6 am-midnight. Door to door. Purchase ticket inside terminal. Suburban Limousine, reserve in advance: 203-377-8294.

Hertz.

P $3.50/day.

BRIDGETOWN, Barbados
Grantley Adams Intl Airport, 9 mi (14.5 km) E

BD$30 ($15) flat rate to places within 9 mi or 30 min of airport, including Bridgetown. To drop-off 30-40 min or 15 mi, BD$30-37 ($15-19). Maximum charge to furthest point of island, North Point, St. Lucy, BD$50 ($25), 55 min. Use only authorized airport taxi. No tip necessary.

Barbados Transport bus boards directly opposite terminal on highway. Every 30 min 6 am-Midnight. BD$1 (50¢), 30 min to city center. Limited baggage only.

Barbados Rent A Car, Dear's Garage, and many others.

P BD$4.50 ($2.25)/day.

BRISBANE, Australia
Brisbane Intl Airport, 7 mi (11 km) NE

A$15 ($12), 25 min.

Skennars bus every 30 min. A$5 ($4), pay driver. 20 min to city.

Avis, Budget, Comcar, Hertz, Thrifty.

P ST: A$12 ($9.60); LT A$10 ($5.80).

BRUSSELS, Belgium
Brussels National Airport, 7.5 mi (12 km) NE

BF1000 ($29), 20-30 min to city center. Metered fare includes tax, tip; make sure meter is on. RT ticket for BF1200 ($35) good for 2 months from Autolux taxis distinguished by small sign in top right corner windshield or back window.

Escalator to lower level entrance for train. Buy ticket BF110 ($3.20) at office. Penalty of BF600 ($17) for boarding train without ticket when office is open. Comfortable, plenty of baggage space. Trains every 20 min from 5:25 am to 11:46 pm. 14 min to Brussels Nord, a Eurail transfer point; 18 min to Brussels Central, end of line.

Ansa, Avis, Budget, Dollar, Europcar, Hertz, Transcar, Venwijk.

P BF220 ($6.40)/day.

BUCHAREST, Romania
Otopeni Intl Airport, 10 mi (16 km) N

City bus with special baggage compartment meets flights. Pay Leu14 (7¢) on boarding. 30-45 min ride to Tarom Airways headquarters in city where taxis are available.

ACR, Avis, ONT Carpati.

BUDAPEST, Hungary
Ferihegy Airport, Terminal 1 (Intl), 10 mi (16 km) SE

HUF500 ($3), 20-30 min to city. Reserve cab at Volan Taxi counter next to rental cars. Flat rate includes service but a small tip will be appreciated.

Malev flights arrive at more distant Terminal 2. Cab: HUF250, 30-40 min to city.

Volan Bus No. 1 to Engels Terrace, downtown. Bus leaves airport from stop at far right outside terminal. Every 30 min 6 am-11:30 pm. HUF20 (40¢) to driver. Good baggage space, comfortable. From Terminal 2, HUF80.

No. 93 from far left as you exit terminal. 4:45 am-11:25 pm at 10-20 min frequencies. Purchase blue ticket from self-service machine inside terminal, on right as you head toward exit. HUF15 (6¢) for two tickets. Validate ticket in machine on bus. No baggage racks, limited floor space. But good chance you'll get a seat because airport is start of the route. To central Budapest, ride to end of bus line, transfer to M3 metro. Fare: HUF1 (2¢) into turnstile. Get off at Deak Terrace downtown. Total travel 50 min. Total cost 5¢! (Also picks up at Terminal 2.)

Avis, Budget, Europcar, Hertz, VIP, and others.

BUENOS AIRES, Argentina

Ezeiza Airport, 23 mi (37 km) SW

Limo-type cars run by Manuel Tienda Leon. $40 center city. Drop you off anyplace, then to downtown bus station. Very comfortable.

Special bus from both terminals to downtown bus terminal, stopping at some hotels. Pay $10 fare before boarding. 30-45 min trip.

Avis, Budget, Hertz, National.

BUFFALO, New York

Buffalo Intl Airport, 8 mi NE

$14, 15-20 min.

ITA Van $5 to downtown hotels.

No. 24A Genesee bus stops at East and West terminals. $1.10. 45-min ride to Lafayette Square.

Alamo, American Intl, Avis, Budget, Hertz, National, Snappy, Thrifty.

P ST $10, LT $6.75, or $27 weekly.

BURBANK, California

Burbank-Glendale-Pasadena Airport, 14 mi NW of Los Angeles

Burbank, $7-9; Glendale, $9-12; Pasadena, $20; downtown LA, $25-26. 25-45 min.

SuperShuttle to Burbank, Glendale, $14-19 first passenger, $8 each addl; Pasadena, $20; downtown L.A., $20; info: 818-244-2700. Also, Prime Time to downtown L.A., $18 first passenger, $7 each addl. Info: 818-901-9901.

No. 94 RTD. Take free Transporter shuttle bus to RTD bus stop on Hollywood Way at airport entrance. Service every 20 min 5:04 am-12:53 am. Approx 1-hr ride to downtown L.A. $1.10.

To LAX SuperShuttle, $15. Reservation a day in advance advised: 818-244-2700.

Alamo, Avis, Budget, Dollar, General, Hertz, National, Thrifty.

P ST $24, LT $13. Remote $7, free shuttle. Valet $18.

BURLINGTON, Vermont

Burlington Intl Airport, 3 mi E

$6.50, 10 min.

Avis, Budget, Dollar, Hertz, National, Thrifty.

CAIRNS, Australia

Cairns Intl Airport, 4.5 mi (7.5 km) N

A$9 ($7.10), 10 min. To northern beach areas, 15-30 min. If no cab available use free phone or dial 515-333. Tip not customary.

Airporter Shuttle Service (Australia Coach) meets flights. A$4 ($3.15) OW, A$7 ($5.55) RT. 10-20 min.

Avis, Budget, Hertz, Thrifty.

P A$4 ($3.15)/day.

CAIRO, Egypt

Cairo Intl Airport, 9 mi (15 km) N

E£12-14 ($5.65-6.60), 40-60 min. Shared rides common, but E£18 ($8.50) will buy private ride during day peak period. Disregard meter—determine fare in advance with driver. Tip E£1-2 (50-95¢). Cabs not usually A/C.

Limo Bank Nasser. To Nile Hilton E£5 ($2.35); Gezira Sheraton E£8.5 ($4); Mena House E£10 ($4.70); Heliopolis E£4 ($1.90). Pay fare aboard. Stops at major hotels, other places on request. A/C. 45-70 min depending on route, traffic. Also goes to pyramids: inquire. Meets all flights.

Avis, Budget, Hertz, InterRent, National.

CALCUTTA, India
Calcutta (Dum Dum) Airport, 17 mi (27.5 km) NE

R70-80 ($4-4.60), 60-70 min to city center.

Airport bus meets flights. R10 (60¢). Since a cab connection may be necessary from city terminal, it may be more convenient to take cab at airport.

Not available in India.

CALGARY, Alberta
Calgary Intl Airport, 5 mi NE

C$15-16 ($12.75-13.60), 30 min downtown.

Airporter every 30 min 6:30 am-11:30 pm to Westin hotel. C$6 ($5) OW, C$11 ($9.35) RT.

No. 57 shuttle bus every 30 min 6:46-8:46 am; 12:16-10:57 pm. C$1.25 ($1), get a transfer. Baggage OK. Change at Whitehorn Station (15 min) to train No. 202 downtown. 14 min to City Hall.

Avis, Budget, Dollar, Hertz, Thrifty, Tilden.

CAMBRIDGE, England
Cambridge Airport, 2 mi (3.2 km) E

£4 ($6.95), 10 min, 10p each add'l passenger, 10p/bag. Tip 10%. Call from arrival hall.

Marshal's. Tel. 62211.

CANBERRA, Australia
Canberra Airport, 3.5 mi (6 km) E

Cabs at airport peak times only. Telephone 285-9222. A$8.50-10 ($6.70-7.90), 6-10 min. Woden, A$13.50 ($10.65), 15 min; Belconnen, A$15.80 ($12.50), 20 min; Tuggeranong, A$22 ($17.40), 25 min.

Avis, Budget, Hertz.

P A$2-4 ($1.55-3.10)/day.

CANCUN, Mexico
Cancun Intl, 12 mi (20 km) SW

P15000 ($5.25), maximum 3 people.

To hotel zone in a combi (minibus), P4000 $1.40 per person for up to 5. 15 min.

Avis, Dollar Turisel, EconoRent Xelha, Hertz, National.

P $3.50/day.

CANTON, Ohio See Akron-Canton, Ohio.

CAPE TOWN, South Africa

D.F. Malan Airport, 15 mi (23 km) NW

Rn36 ($13.50), 20 min.

Interkaap bus meets every flight. Pay Rn8 ($3) aboard. Stops at South African Airways terminal in Cape Town. 25 min.

Avis, Budget, Imperial.

CARACAS, Venezuela

Simon Bolivar (Maiquetia) Intl Airport, 13.6 mi (22 km) NW

Buy trip ticket at airport authorized taxi booth, $5, 35-40 min. Some airlines provide free transfer; inquire.

Caracas Aerop. Internacional coach hourly 5:30 am-12 midnight. B40 (80¢), 30 min. Stops at metro station and Central Park.

Catia La Mar - Caracas bus hourly 6 am-6 pm. B20 (40¢), 45 min to city center.

Budget, Hertz, National.

P B42 (85¢)/day.

CASABLANCA, Morocco

Nouasseur Airport, 22 mi (35 km) S

Dh100 ($15), 30 min. Tip $2.

CTM coach hourly on the hour from front of terminal. 40 min to city bus terminal.

Avis, Budget, Europcar, Hertz, Interloc, InterRent, Saudia, Safloc.

P $2/day.

CASPER, Wyoming
Natrona County Intl Airport, 8 mi NW

$8, 20 min; extra passengers, $2 each.

Avis, Budget, Hertz, National.

P LT $2.75.

CAYMAN BRAC, Cayman Islands
Gerrard-Smith Airport, 5 mi W

CI$6 flat rate, 10 min. Tip 10%.

Avis.

CEDAR RAPIDS, Iowa
Cedar Rapids Municipal Airport, 8 mi S

$11. 10-15 min to Cedar Rapids.

DJs Limousine, $7-9 to Cedar Rapids; $10 OW, $18 RT to Iowa City. Look for desk at baggage claim. Meets flights.

Avis, Budget, Hertz, National.

CHAMPAIGN/URBANA, Illinois
Univ. of Illinois-Willard Airport, 6.5 mi SW

$8-10, 20 min. Two or more passengers, $6 each.

Avis, Budget, Hertz, National.

P $5/day.

CHARLESTON, South Carolina
Charleston Intl Airport, 12 mi NW

$18. 20-25 min.

$7, 25 min, departs every 15 min.

Alamo, American Intl, Avis, Budget, Dollar, Hertz, National, Thrifty.

CHARLESTON, West Virginia
Kanawha Airport, 4 mi NE

$7, 10-15 min.

Avis, Budget, Hertz, National.

CHARLOTTE, North Carolina
Charlotte-Douglas Intl Airport, 8 mi W

$11. Bags 50¢. 15 min.

Yellow Cab Limousine, $4. 20-25 min to downtown hotels. Operates 7 am-11 pm. Call for Sat-Sun service: 704-332-6161.

Avis, Budget, Dollar, General, Hertz, National.

P ST $10, LT $3.

CHATTANOOGA, Tennessee
Lovell Field, 10 mi E

$10 flat rate per person, 20-30 min.

No. 19 bus every 55 min. 75¢.

Avis, Budget, Dollar, Hertz, National, Thrifty.

CHICAGO, Illinois
Meigs Field (lakefront), 2.3 mi SE of First Natl Bank Bldg in Loop

$5-7, 5-10 min to Loop, N. Michigan Ave. Cabs usually available weekdays. If not use cab phone at Butler Aviation desk. (Incoming flights about 10-15 min from touchdown can use Butler intercom frequency 122.95 to request cab.)

Avis.

P Free. Limit 2 hrs.

CHICAGO, Illinois
Midway Airport, 10 mi SW

$17-19. Share-Ride $10 to Loop. 20-40 min.

Airport Express vans every 15 min 8 am-9:30 pm. $9 OW, $16 RT. To Hilton, Palmer House, Marriott, Hyatt Regency, Westin, other hotels. 20-50 min. Res required for return: 312-454-7799. **South Suburbs** Tri-State Coach at 6, 7 am; hourly at :45 to 8:45 pm; 9:10, 10:40, 12:10 am. To Oak Lawn, Alsip, Harvey, Homewood, Matteson, Lansing. $10. Also Indiana, Michigan connections. Info: 800-248-8747. **To LaPorte, Michigan City, Michiana Airport South Bend, Notre Dame, Mishawaka, Osceola, Elkhart** United Limo. Info: 800-833-5555.

No. 99M M-F rush hour bus picks up along Cicero Ave, just outside terminal. Service at 12-14 min intervals 6 am-8:15 am inbound, 4 pm-6 pm outbound. Via Stevenson Expy to State St Mall, downtown. $1.15. Often crowded and perhaps not best choice for travellers with much luggage. Transit to Midway will take great leap forward when rail link is finished in 1992.

To O'Hare C. W. Airport Service hourly 7 am-9:45 pm intervals. $9.75. 45 min. Info: 312-493-2700.

Alamo, Avis, Budget, Dollar, Hertz, National, Thrifty.

P ST $15, LT $6.

CHICAGO, Illinois

O'Hare Intl Airport, 18 mi NW

$24. Addl passengers 50¢ each. 25-60 min downtown depending on time of day. Share-a-Cab $15. **To Midway Airport** $32-35.

Airport Express to downtown hotels $12.50 OW, $22 RT. Every 10-30 min 6:30 am-11:30 pm. **To North Shore Suburbs** Airport Express door-to-door to Evanston, Glencoe, Glenview, Highland Park, Kenilworth, Lincolnwood, Northbrook, Northfield, Skokie, Wilmette, Winnetka. Departs lower level outside baggage claim. Res & info: 312-454-7799. **To South Suburbs, NW Indiana, SW Michigan** Tri-State Coach from lower level at United and between terminals 2 and 3. Hourly 6:35 am-8:35 pm to Harvey, Homewood, Matteson. To Harvey only at 10:15 pm, 11:45 pm, 12:35 am. $12, 45-70 min. To Hammond, Glen Park, Merrillville hourly 5:45 am-8:45 pm, also 10:15, 11:45 pm. $14. Connection to Stevensville, New Buffalo, Mich. Info 312-374-7200.

CTA subway-elevated train from terminal inside O'Hare direct to Loop (downtown) in 35-40 min. From baggage claim, follow rail-car signs down escalator or elevator to lower level, where moving sidewalks converge at station entrance. $1.25 fare to agent or exact coins in turnstile. Trains every 3-5 min peak hours, up to 30 min late night. 24-hr service. Baggage racks on some cars, some floor space. May be tight squeeze in rush hours. Trains, stations generally clean. Loop platforms crowded in rush hours and could be better lighted. Travelers with little luggage and bound for central business district will find the O'Hare subway an excellent choice. During rush hours it often makes better time than cabs, limos. Airline employees are among regular riders.

To Midway C.W. Limo hourly 6:30 am-1:30 pm, 2:15-5:15 pm, 6-10 pm, last at 10:45 pm. $9.75, 45 min. Info: 312-493-2700.

Airways, Avis, Budget, Dollar, Hertz, National.

P ST $16, LT $6. Free shuttle.

All the following board from lower level between terminals 2 & 3: **To Hyde Park (U. of Chicago)** C.W. Limo—See schedule above "To Midway." $9.75.

Stops at Hyde Park Hilton, Windermere, Del Prado, Intl House, Ida Noyes Hall. 312-493-2700. **To Madison, Janesville, Beloit** Van Galder Alco 800-747-0994. Greyhound 608-257-0440. **To South Beloit, Janesville, Madison (Wis)** Van Galder Alco 800-747-0994. **To Lake Geneva** Owl Airport Bus Service 312-427-3102. **To Rockford, Peoria** Peoria-Rockford Bus Co. 815-963-3424 **To Joliet, Normal, Peoria.** Peoria Charter Coach 800-448-0572. **To Champaign, Urbana, U. of Illinois, Rantoul** O'Hare Shuttle 800-642-7388. **To LaPorte, South Bend, Mishawaka, Elkhart, Ind.** United Limo 800-833-5555 (US) **To Deerfield, Lake Forest, Waukegan** Airport Express 312-454-7800.

To Mitchell Field, Milwaukee Greyhound from Terminal 1 at 12:35, 5:30, 11:30 am; 1:30, 3, 5:45, 8 pm. $13 OW, $20 RT. 1 hr 55 min trip. Info: 414-272-2954.

CHIHUAHUA, Mexico

Genl Roberto Fierro Villalobos Airport, 11 mi (17.6 km) NE

To Center City, P2950 ($1); residences, P8150 $2.85.

Ansa, Avis, Budget, Fast, Hertz, National.

P $6.30/day.

CHRISTCHURCH, New Zealand

Christchurch Intl Airport, 7 mi (11 km) NW

NZ$15 ($8.70), 15 min. No tip.

Canride to city center, hourly from 6 am to 8 pm. NZ$2.40 ($1.40)

Airport bus about every 30 min 6:35 am-10:15 pm M-F.Sat from 7:17 am. Limited service Sun, Good Friday, Christmas Day. 30 min to city, NZ$3 ($1.75).

Avis, Budget, Hertz.

P NZ $10 ($5.90)/day.

CINCINNATI, Ohio

Greater Cincinnati Intl Airport, 13 mi SW

$20. 20-30 min.

Jet Port Express every 30 min 5:30 am-11 pm Sun-Fri; hourly Sat 5:30 am-10:30 pm. To Westin, Hilton, Cincinnatian, Clarion, Hyatt Regency, Netherland Plaza. $8 OW, $12 RT. 30 min to city.

Alamo, Avis, Budget, Dollar, Hertz, National, Thrifty.

P ST $5.50, LT $2.75

CLEVELAND, Ohio

Cleveland-Hopkins Intl Airport, 10 mi SW

$16-17, 15-30 min to Public Square, downtown. To Cleveland Clinic, $22, 30 min.

Hopkins Airport Limo to Painesville, Mentor, Willoughby, Wickliffe, Euclid, North Randall, Shaker Heights, Beachwood, Solon, Wilson Mills, Oberlin, Elyria, Kent U., and vicinity. Frequent departures. Fares range $9-$12. Departs from Exit Door 6. Info: 216-267-8282.

Airport/Windermere train from station inside air terminal every 11 min rush hours, 15-20 min other times from 4:30 am to 1 am M-F. Runs at 15-30 min intervals Sat-Sun-Hol. $1. 22 min to Cleveland Union Terminal (Public Square), downtown. Clean, bright coaches and stations, ample baggage space.

To Akron-Canton Airport Hopkins Airport Limo. Scheduled service is provided during the hours of 7 am-11:45 pm M-F, stopping at Residence Inn, Radisson, Comfort Inn, Holiday Inn, Hilton Fairlawn, Holiday Cascade, Quaker Sq. Hilton. $12.75. Info: 216-267-8282.

Alamo, Avis, Budget, Dollar, Hertz, National, Snappy, Thrifty.

P ST $14, LT $8.

COLOGNE/BONN, Germany

Cologne/Bonn Airport, 8.7 mi (14 km) SE of Cologne, 12 mi (20 km) NE of Bonn

To Cologne, DM33 ($19), 20 min; to Bonn, DM52 ($30), 25 min. To Aachen, DM160 ($93), 1 hr; to Wuppertal, DM125 ($72.50), 1 hr; to Leverkusen, DM48 ($28), 35 min. Tip 10%.

To Cologne, Line 170 coach from arrival level departs at 20-30 min intervals 6:35 am-11:35 pm, 20 min to city. Pay driver DM3.60 ($2.10). To Bonn, Bonn Line FL departs every 30 min 5:40 am-11 pm, 25 min to city. Pay driver DM7.20 ($4.20).

Autohansa, Avis, Europcar, Hertz, InterRent AG, Sixt.

P ST DM10 ($5.80), LT DM5.20 ($3). In the Parkplatz Nord lot, parking is free for passengers departing on holidays of at least 1 week.

COLOMBO, Sri Lanka
Katunoyake Intl Airport, 20 mi (32 km) N

Rs550 ($13.40), 50 min. Buy ticket at taxi counter. Look for "Airport Taxi" sticker on sunshade. To Kandy (Hill capital), Rs1600 ($39), 2 hrs; to Anuradhapura (ancient city), Rs2817 ($68.70), 5 hrs; to Hikkaduwa beach resort), Rs1943 ($47.15), 3.5 hrs. No tip expected but Rs10-20 will be appreciated on Colombo run, more on longer jaunts.

Coach leaves from main exit opposite terminal every 30 min from dawn to late night. Rs5.50 (15¢), 60 min to main bus stop in Colombo.

Sri Lanka Transport Board operates service to Kandy: Rs25 (50¢), 3 hrs; and Negombo: Rs4 (10¢), 20 min. Inquire at taxi counter.

Avis, Europcar, Hertz.

P Rs10 (25¢)/day.

COLORADO SPRINGS, Colorado
Colorado Springs Municipal Airport, 8 mi SE

$10 downtown, Broadmoor Hotel $14. 15-20 min.

Avis, Budget, Dollar, Enterprise, Hertz, National, Payless, Thrifty.

P ST $7.50, LT $4.

COLUMBIA, South Carolina
Columbia Metropolitan Airport, 6 mi SW

$10 for one or two, 50¢ each addl passenger. 10-15 min.

Alamo, Avis, Budget, Dollar, General, Hertz, National, Payless.

COLUMBUS, Georgia
Columbus Metropolitan Airport, 5 mi NE

$8, 10 min. To Ft. Benning, $17, 20 min. Extra passengers, 50¢ each. Baggage, 50¢ each.

Avis, Budget, Hertz, National.

P ST $6/day, LT $4/day.

COLUMBUS, Ohio
Port Columbus Intl Airport, 7 mi NE

$15, 20 min.

Airport Shuttle every 30 min 8 am-11 pm. Sun from noon. $6.50 OW, $12 RT. To Holiday Inn, Christopher, Hyatt, Hyatt Capitol, Great Southern, bus station. Info & res: 614-478-3000.

Avis, Budget, Dollar, Hertz, National.

P ST $16.25, LT $8.25, Remote $5.50/day, $33/week.

COPENHAGEN, Denmark

Copenhagen Airport, 6.25 mi (10 km) SE

DKr90 ($14), tip incl; 15-20 min. Metered taxis with signs atop.

SAS bus, DKr26 ($3.90), from outside arrival hall to Hotel Scandinavia, Central RR Stn. Departures every 15 min from 5:45 am. Last at 11:10 pm. Baggage racks. 20-30 min ride.

Outside arrivals hall. Pay DKr14 ($2.10) when boarding. Departures every 10-20 min 4:31 am-11:48 pm. 30-45 min to Town Hall Square. Little baggage space, 20 stops into city. First bus Sun 5:22 am.

To Malmo, Sweden and other Swedish coastal cities, 6 am-8 pm. Ask Airport Information.

Avis, Budget, Europcar, Hertz, InterRent, Pitzner.

P ST DKr90 ($13.50), LT DKr60 ($9), 8 days DKr305 ($45.75).

COZUMEL ISLAND, Mexico

Cozumel Airport, 1.9 mi (3 km) N

$3, 5-10 min.

Servicio Colectivo from front of terminal every 10 min 7 am to last flight. 10 min, 75¢.

Avis, Budget, Hertz.

DALLAS, Texas

Dallas Love Field, 7 mi NW

$12, addl passengers $1 each. 15-20 min.

SuperShuttle vans to downtown, Market Center hotels upon contacting by courtesy phone in baggage claim, $8.50. Van to DFW Airport on the hour, $13.50.

DART No. 39 Love Field bus at 15 min intervals peak times, 30-40 min other times. 5:28 am-10:31 pm. 75¢ exact. Approx 35 min to Lamar & Main, downtown. Take return bus along Commerce St.

Avis, Budget, Dollar, General, Hertz, National.

P ST $6, LT $4.

DALLAS-FT. WORTH, Texas

Dallas-Ft. Worth Airport, 21 mi NW of Dallas, 24 mi NE of Ft. Worth

To Dallas $27-28 plus extras: $1 airport toll; $1 each addl passenger, toll road. Transfer between terminals, $5. Minimum fare leaving airport $10. 25-40 min to downtown Dallas. **To Ft. Worth** $31-32, same addl charges as above. 30-40 min.

SuperShuttle vans to downtown Dallas and Market Center hotels. On call 24 hrs. $10.50 to hotels, $13.50 to other city addresses. Collect baggage, call dispatch using free phone at ground transportation board or 817-329-2020. Fares to other points: $11.50-30.50. Other van downtown: on call 24 hrs, $10. Others: Best Shuttle (214-244-4306), Big D Shuttle (214-750-9842), Discount Shuttle (214-445-1441), Limaxi (214-826-6997), Liner (817-467-4718), On Time Shuttle (214-980-1995), RBS (214-631-4441). **To Ft. Worth** T Airporter at 5:42, 6:42, then every 30 min to 9:10 pm, then at 10:21 pm, 11:21 pm. $7. To Airporter Park & Ride, Worthington, Hyatt, Chisholm. AMEX, MC, VISA. Info: 817-334-0092. Also SuperShuttle. Hotels $12.50. AMEX, MC, VISA. TBS $12.

American Airlines Park-and-Ride AAirlink from Dallas parking lot at LBJ Fwy (I-635) & Midway Rd. Park free, take SuperShuttle to DFW. $11.50. 30-40 min.

To Love Field SuperShuttle $13.50 first psgr, $4.50 ea addl. Call dispatch using free phone at ground transportation board.

Avis, Budget, Hertz, National. No agents in baggage claim–ride AIRTRANS or courtesy car to rental service areas in north and south parking lots. Also on call: Alamo, American Intl, Dollar, General Thrifty, others.

P ST $12, 9, LT $5, 7. Free shuttle.

To Waco The Streak at 11 am, 2, 5, 8 pm. $25 OW, $45 RT to Hillsboro; $30 OW, $55 RT to Waco hotels, Baylor. **To Wichita Falls** Skylark Van at 9:30 am, 12:30, 3:30, 5:30, 8:30 pm. $28/$50. Res: 817-322-1352.

DARWIN, Australia

Darwin Airport, 11 mi (18 km) NE

A$13 ($10.25), 15 min. To Nightcliff, A$13 ($10.25), 15 min. Tip not customary but A$1 OK.

Darwin Airport Services meets flights, drops off at hotels, motels throughout city. A$5 ($3.95).

Avis, Budget, Hertz, Territory Truck, Thrifty.

P A$2 ($1.60)/day.

DAYTON, Ohio

Dayton Intl Airport, 12.5 mi N

$18. 15-20 min.

Charter Van every 40 min 5:45 am-11:25 pm. $13 OW, $23 RT. 20-25 min to Marriott, Dayton Plaza, Hilton, Belton Inn, Ramada, Stouffer Inn. Weekends by reservation: 513-898-7171.

Alamo, Avis, Budget, Dollar, Hertz, National.

P ST $10, LT $4.50.

DAYTONA BEACH, Florida

Daytona Beach Regional Airport, 3 mi SW

$10, $1 each addl rider. 15-20 min.

Alamo, Avis, Budget, Hertz, National.

DELHI, India

Palam Airport, 10 mi (16 km) W

R60 ($3.50), 40-45 min. Atel cabs, yellow and black, are metered. Make sure meter is running. No tip expected. Baggage R5-10 (30-60¢) apiece. Not A/C. For A/C comfort, inquire at Ashok Travel and Tour desk to reserve car and driver. R134-168 ($7.60-9.60) flat rates. 30-40 min into city.

Not available in India.

DENVER, Colorado

Stapleton Intl Airport, 7 mi NE

Toll free US number for Denver airport ground transportation info: 800-AIR-2-DEN. Also 303-270-1750.

Downtown hotels, $10-14, 15-20 min. 40¢ each addl passenger. Tech Center $16-23; Aurora area $8-18; Lakewood-West $19-23; Lakewood-Southwest $20-22; Northglenn $22-24; Littleton $21-27; Englewood $17-19; Boulder $40-47. Ski areas: Steamboat $204; Winter Park $94; Vail $132; Aspen $264.

Van every 15 min M-F, 30-45 min Sat-Sun, to Brown Palace, Radisson, Hotel Denver, Holiday Inn, Fairmont, Executive Towers, Westin, Embassy Suites, Marriott. $6 and Denver Airport Limo, $5. 20-25 min downtown. Buy ticket at Airporter counter in Ground Transportation Center, lower level opposite Door 6. **To Boulder** Boulder Airporter hourly 9 am-10:15 pm. 40-min run to Boulderado Hotel, Clarion, Broker Inn. $9.50, and Supercoach, $9.

To other Denver-area locations Call toll-free 800-AIR-2-DEN or inquire at Ground Transportation Center. Many scheduled and unscheduled services link Stapleton Airport and points in metro Denver; Ft. Collins; Cheyenne and Laramie, Wyoming; Scotts Bluff, Nebraska; Vail and other ski/resort areas; military installations.

No. AB express bus to Denver, Boulder. 7:10, 8:10 am then every 30 min to 2:40 pm; then 3:12, 3:37, 4:07, 4:37, 5:07, 5:52 pm; then hourly 6:40 pm-12:40 am. Sat hourly from 6:40 am. Sun-Hol hourly from 7:40 am. Fare $1.50 Denver, $2.50 Boulder. Baggage OK, bicycles if space available. 20 min to Denver Bus Center (20th & Curtis); 80 min to Boulder Transit Center (14th & Walnut). Also No. 32 bus 4:51 am-12:50 am M-F at 15-min intervals 6-9 am, 30 min other times. Board at Red Zone, outside Door 12. $1 exact. 23 min to 17th & California, downtown. Operates 5 am-12:44 am Sat-Sun at 30-min intervals.

Alamo, Avis, Budget, Dollar, General, Hertz, National, and others.

P First 30 min free all lots. Then ST $8.50, LT $4. SMART shuttle all hours every 10 min.

DES MOINES, Iowa

Des Moines Intl Airport, 5.5 mi SW

$10 to Capitol, 50¢ each addl rider. 15-20 min.

Avis, Budget, Dollar, Hertz, National.

P ST $10, LT $4/$3.

DETROIT, Michigan

Detroit City Airport, 6.5 mi NE

$11, 15-30 min. To Metro airport, $30, 30-45 min.

Avis, Hertz, National.

P ST $9. LT $5, $25/week.

DETROIT, Michigan
Detroit Metro. Wayne County Airport, 19 mi SW

$28, 30-45 min to downtown Detroit. Romulus $10.20; Southfield $30.80; East Detroit $37.50; Ann Arbor $34.80; Dearborn $18.40; Pontiac $41.90.

Commuter Transportation Co. vans every 30 min 7:15 am-midnight M-F, hourly on the hour Sat-Sun. $12 OW, $22 RT. To Westin, Omni, Cobo Hall, Leland, Pontchartrain, Days Inn. **To New Center & Dearborn** Hourly 7:15 am-11:15 pm to Marriott, Holiday Inn, Ritz Carlton, Hyatt Regency, Best Western, St. Regis. $10-15 OW. **To Southfield** Hourly 6:45 am-midnight. $15.50 OW, $29 RT. Also service to: Grosse Pointe, Fraser, Warren (GM Tech Center), Roseville, Farmington Hills, Novi, Livonia, Plymouth, Rochester, Troy, Hazel Park, Ypsilanti, Ann Arbor. Info & Res: 800-458-9401 (US), 800-351-LIMO (Mich). AMEX, MC, Visa. **To Toledo** JetPort Express at 6:55, 7:55, 9:10, 11:10 am; 1:10, 3:10, 4:40, 6:10, 7:40, 9:10, 10:40 pm; 12:25 am. To Holiday Inn Monroe, Sheraton Westgate, MCO/Hilton, Holiday Inn Southwyck, Radisson, Holiday Inn Riverview, Marriott. $12 OW, $20 RT. Info: 800-543-9457.

To Detroit City Airport Commuter Transportation Co. vans hourly on the hour 7 am-midnight. $12.50. 30-45 min.

Alamo, Avis, Budget, Dollar, Hertz, National, Thrifty.

P ST $8, LT $4.50, $24 weekly. Free shuttle every 10-15 min.

DRESDEN, Germany
Dresden Airport, 6 mi (10 km) N

To Hotel Bellevue, city center, DM30 ($18), 20 min. Tip: round up fair to nearest mark or 50 pfennig.

Airport bus leaves 30 min after flight arrivals. Arrives Central RR stn in about 30 min.

Avis, call 58-31-41. Hertz at Bellevue Hotel, call 56-620.

DUBAI, United Arab Emirates
Dubai Intl Airport, 10 mi (16 km) N

Di30 ($8), pay in advance at taxi desk. Bags Di2 (55¢) each. Pleasant 15-20 min ride on oceanside road. To Abu Dhabi Di200-250 ($55-70).

Avis, National.

DUBLIN, Ireland
Dublin Airport, 8 mi (13 km) N

IR£7.50 ($11.85), 20-30 min to city center. Extra riders 40p (60¢) ea. Baggage 40p (60¢) ea. Tip 10%.

Airport Service from Arrivals Terminal at 20 min intervals 8 am-9:30 pm. Pay driver IR£2.30 ($3.60). Approx 25 min to Gresham Hotel, Store St. bus terminal.

No. 41A boards at bus stop, a short walk from arrivals hall. Departures about every 30 min 7 am-11:20 pm. Pay driver 90p ($1.40). Double-deckers have small amount of storage space beneath stairs. 30 min to Eden Quay, many stops en route. Buses No. 41, 41C also at peak times. For traveler with little luggage, this can be a delightful top-deck trip into town.

Avis, Budget, Europcar, Hertz, Johnson & Perrott, Murrays.

P ST IR£5 ($7.90), LT IR£2.50 ($3.95).

DUNEDIN, New Zealand

Dunedin Airport, 18.6 mi (30 km) SW

NZ$35 ($20), 30-35 min. Cabs not always at airport; use free taxi phone. To Port Chalmers Wharf NZ$44 ($25.50), 60 min; Queenstown NZ$320 ($185), 3 hrs; Alexandra NZ$210 ($122), 2 hrs; Invercargill NZ$247 ($143), 2 hrs 10 min. No tip.

Several different minivans meet flights and provide to-your-door service, NZ$10 ($5.85), 30-45 min.

Avis, Budget, Hertz.

DURHAM, North Carolina See RALEIGH/DURHAM, North Carolina

DUSSELDORF, Germany

Dusseldorf Airport, 5 mi (8 km) N

DM20 ($11.60), 15 min. Tip 5%.

Bus 727 to Hauptbahnhof (main central station) every 20-30 min 5:21 am-midnight (from 5:41 am Sat, 5:49 am Sun). Boards on arrivals level outside taxi ranks. Comfortable, plenty of baggage space. DM2.60 ($1.50). 25 min.

S-Bahn No. 7. Follow signs from baggage claim to lower-level terminal. Long tunnel connects to rail station but luggage conveyor belt makes transfer somewhat easier. Self-service ticket machine dispenses zone fare cards. Two zones to Hauptbahnhof, downtown, costs DM2.60 ($1.50). Tickets must be cancelled at validating machine before boarding. Trains every 20 min 5:03 am -12:33 am. Plenty of luggage space, A/C. 13-min ride.

Autohansa, Avis, Europcar, Eurorent, Hertz, InterRent, Sixt-Budget.

P ST DM12.50-15 ($7.50-9), LT DM4.50 ($7.80) depending on location. There are 10 garages or lots with a total capacity of 10,000.

EDINBURGH, Scotland
Edinburgh Airport, 7 mi (11 km) W

£12 ($20.40), 15-25 min. Cabs say Airport Taxi. They are metered and take up to four people. No extra charges. Tip 15%. Look for taxi rank near Door D between gates 3 and 4.

No scheduled service but Royal Scot Hotel operates courtesy coach.

Airlink No. 100 departs from bus stop outside Door G. No special baggage space. Departs every 30 min 7 am-10:30 pm except Sat-Sun when service is hourly. No service Christmas, New Year's days. Pay £2.50 ($4.35) on bus. Stops at Ingliston, Maybury/Royal Scot Hotel, Drum Brae, Zoo, Murrayfield, Haymarket, West End, Waverley Station. 25-30 min ride.

Alamo, Avis, Eurodollar, Europcar, Hertz.

P £4.10 ($7.15)/day.

EDMONTON, Alberta
Edmonton Intl Airport, 20 mi (32 km) S

C$32 ($28.15), 30 min.

Grey Goose Airporter every 30 min 6:30 am-1 am M-F. Hourly Sat-Sun. C$11 (9.70) OW, C$18 ($15.85) RT. Serves Westin, Four Seasons, Chateau Lacombe, Sheraton, and other hotels.

Avis, Budget, Dollar, Hertz, Thrifty, Tilden.

P ST C$5.90 ($5.20)/day.

EL PASO, Texas
El Paso Intl Airport, 5 mi E

$14, 50¢ each addl rider. 15-20 min.

Sprint van every 15 min. $9 downtown, $10 Ft. Bliss, $12 West Side.

Avis, Budget, Dollar, Hertz, National.

EVANSVILLE, Indiana
Evansville Regional Airport, 6 mi N

$10-11. 10 min.

Avis, Budget, Dollar, Hertz, National.

P ST $8, LT $3.50.

FAIRBANKS, Alaska
Fairbanks Intl Airport, 7 mi W

$7.50, addl passengers $1 each. 15 min.

Most hotels provide a courtesy van or will reimburse cab fare.

Allstar, Arctic, Avis, Budget, Hertz, National, Payless.

P ST $6/day.

FARGO, North Dakota
Hector Intl Airport, 3 mi NW

Cabs do not wait at airport. Look for phone in vestibule. $7-10, 10 min. To hotels in S. Fargo, $7-10, 10-15 min. Extra passengers 50¢ ea.

Avis, Budget, Hertz, National, Thrifty.

P ST $5.50/day, LT $4/day.

FLINT, Michigan
Bishop Intl Airport, 5 mi SW

$8, $2 each addl. 15 min.

Avis, Dollar, Hertz.

P ST $3.25/day.

FLORENCE, Italy
Peretola Airport, 3.1 mi (5 km) NW

Lit2500 ($2), 15 min. If fare is fixed, no tip. Otherwise tip 5-10%.

Avis, Hertz.

FORT-DE-FRANCE, Martinique
Lamentin Airport, 4.4 mi (7 km) ESE

60F ($10), 10 min. Or take a collective taxi, identified with TC sign on roof, for F4.1 (80¢). To Pointe du Bout resorts, F142 ($23.50), 30 min; Leyritz Plantation, 273F ($45.50), 75 min; Ste-Anne, 252F ($42), 60 min. If driver accepts a pet he can charge F3 (50¢). Tip 10%. 40% surcharge 8 pm-6 am.

Scenic alternative Take cab from airport to ferry dock: 60F ($10). Catch ferry across bay: 12F ($2.00). Departures at 6:30, 7:30, 8, 8:30, 9, 10, 11, 11:45 am; 12:15, 12:45, 1:45, 2:30, 3, 4, 4:30, 5, 5:30, 6, 7, 10, 11:15 pm; 12:10 am. 20-min ferry crossing to town, landing adjacent to Meridien, Le Bakoua, other hotels. Much less than the price of a cab and great fun if you're traveling light.

Avis, Budget, Carib, Dollar, Europcar, Hertz, InterRent, Milleville.

P F109 ($22)/day.

FRANKFORT, Kentucky See LEXINGTON/FRANKFORT, Kentucky

FRANKFURT, Germany

Frankfurt-Main Airport, 7 mi (12 km) SW

DM35 ($21), 20 min to central points. Tip 5%.

Airport has its own modern rail station, reached by escalators and elevators. To Frankfurt, buy ticket DM3.60 ($2.15) from machine in Hall B, baggage claim. Note that passengers without tickets pay DM60 ($36) fine aboard. Trains every 10 min. 11-min ride to Frankfurt Hauptbahnhof. Very comfortable. Ample floorspace for bags. Direct rail connections from airport to 27 cities in West Germany and major cities in Austria, Hungary, and the Netherlands, including Wurzburg, Augsburg, Munich, Vienna, Salzburg, Amsterdam, Dusseldorf, Cologne, Basel, Hamburg.

Lufthansa Airport Express This supertrain "flies" 4 times daily along the scenic Rhine River between Frankfurt and Dusseldorf airports with intermediate stops in Bonn, Cologne Central Stn, Cologne-Deutz, and Dusseldorf Central Stn. Only passengers with valid plane ticket may use this service. Luggage automatically transferred from plane to train. Check-in: basement level "Unterm Flughafen" Area B, Counter E-99. Departure times can be found in Lufthansa timetable. Frankfurt to Dusseldorf, 2 hrs 40 min. Also "flies" non-stop to Stuttgart 4 times daily, 90 min.

Lufthansa Airport Bus Specially designed, deluxe, air-conditioned buses with Lufthansa "flight" numbers LH9901 to LH9954 at Arrival Level B in front of the Meeting Point. Reservations and ticketing should be done in advance and may be included with air ticketing; in exceptional cases and after hours the driver can issue tickets. One-way DM33 ($20), round-trip DM55 ($32). **To Heidelberg** (Penta Hotel) every 30-60 min from 7:15 am to 9:15 pm; return to Frankfurt-Main Airport every 30-60 min from 6:15 am to 8:15 pm. 60 min ride time (some buses take 90 min, with one stop in Mannheim). **To Mannheim** (Holiday Inn) every 30-60 min from 7:00 am to 10:45 pm; return to airport every 30-60 min from 6:45 am to 7:30 pm. 60 min ride time. Passengers may check in and their baggage checked through at the above hotels if they are departing Frankfurt on Lufthansa, Condor and other carriers where Lufthansa is the handling agent at Frankfurt. Passengers of other airlines may use the bus but cannot check in for flights at the hotel bus departure points. The hotel counters are open from 6:00 am to 6:00 pm daily, 6:00 am to 1:00 pm Saturday, and 11:00 am to 6:00 pm Sunday.

To Heidelberg TLS Taxi & Limo Service from "Meeting Point," Hall B. DM40 ($23). 24-hr service but reservation required. Phone: 6221-100 99. Fax: 6221-16 27 80.

AK Autovermeitung, Avis, Europcar, Eurorent, Hertz, InterRent, Schuldt/ Autohansa, Sixt/Budget. (Rental counters are on arrival level, Terminal A.)

P DM20 ($12) first 7 days per day; thereafter DM10 ($6) per day. There are more than 10,000 enclosed spaces. Special parking for handicapped drivers available (for information call 69-690-6889).

FREEPORT, Grand Bahama Island

Freeport Intl Airport, 2 mi (3.2 km) NW

$2.50. 5-10 min. Tip 10%.

Avis, National.

FRESNO, California

Fresno Air Terminal, 5 mi (8 km) NE

$12-13, 15 min.

To Coalinga Coalinga Transit, $7 RT. Info: 209-935-1535.

No. 26 and 39 bus outside airport entrance every 30 min 6:42 am-6:15 pm. 75¢.

Avis, Budget, Dollar, Hertz, National.

FT. LAUDERDALE, Florida

Ft. Lauderdale-Hollywood Intl Airport, 4 mi S

To downtown business area $8, beach hotels $11, Galt Ocean Mile $15.

Yellow Airport Limousine, $8-10 downtown, $6-8 beach hotels.

BCt, east end of Terminals 1 & 3, 20 min, 24 hrs a day, 85¢.

Ajax, Alamo, American Intl, Avis, Budget, Dollar, Enterprise, General, Hertz, National, Payless, Snappy, Thrifty.

P ST $7, LT $5/$3.50.

FT. MYERS, Florida
Southwest Regional Airport, 15 mi SE

Midtown $17, 25 min. To Sanibel Island $30, 75 min; Naples, $30-40, 75 min; Marco Is., $60, 75 min.

Sun Lines to Ft. Myers $7.50 each, minimum two.

Express Bus to Naples, Marco at 9:30, 10:30, 11:30 am; 1, 2, 3, 4:30, 5:30, 8, 9, 11:30 pm. 6:30 pm run to Naples only. Naples $8, Marco $12.

Alamo, Avis, Budget, Dollar, Hertz, National, Thrifty, Value, and others.

P ST $8, LT $4.50/day.

FT. WAYNE, Indiana
Baer Field, 9 mi SW

$14, 15-25 min.

Avis, Budget, Hertz, National.

FT. WORTH, Texas See DALLAS-FT. WORTH, Texas

FUNCHAL, Madeira Islands
Funchal Airport, 14.5 mi (23 km) NE

Esc500 ($3.75). 35 min. Tip 10%.

Hotels Casino Park, Madeira Palace, Reids, Savoy, Sheraton operate courtesy vans.

Avis, Hertz, Rodavante.

P Free.

GENEVA, Switzerland
Cointrin Airport, 2.5 mi (4 km) NW

SF25 ($17.25) to city center. Metered, tip usually included. 15 min.

6-min rail link between air terminal and Cornavin RR Station in downtown Geneva. Departures about every 10 min 5:39 am-11:36 pm. Fares: 1st Cl, SF6.40 ($4.40), 2nd Cl, SF4 ($2.75). Plenty of baggage space. Also from Cointrin Airport there are direct rail connections to Neuchatel, Delemont, and Basle in the north; Berne, Lucerne, Zurich, St. Gall in the east; Sion, Brigue, northern Italy to the south via Lausanne. Check baggage through with "Baggage Fly" service for SF9 ($6.20). Ask Swiss Natl Railroads agent.

TPG (Transports Publics Genevois) No. 10 every 5-15 min 5:35 am-12:09 am. To Bel-Air (city center) and Cornavin Station. Picks up on departure level.

Ski season Direct buses from airport to French ski resorts—see Thomas Cook European Timetable.

AI-ANSA, Avis, Europcar, Hertz, InterRent, Budget.

P ST SF20 ($13.60), LT SF8 ($5.50), some free.

GENOA, Italy
Cristoforo Colombo Airport, 4.5 mi (7 km) W

Lit20,000 ($16), 20 min. Tip 5-10%.

AMT bus to main station, 30 min. Lit3500 ($2.80).

Avis, Budget, Eurodollar, Europcar, Hertz, InterRent, Maggiore.

P Lit10,000 ($8)/day.

GEORGETOWN, Grand Cayman
Owen Roberts Airport, 1 mi (1.6 km) SE

CI$5, 5 min. Tip 15%.

Avis, Coconut, Hertz, Holiday Payless, National.

GIBRALTAR, Gibraltar
Gibraltar/North Front Airport, 1/2 mi (0.5 km) N

£1.5 ($2.50) for 1-2 passengers; 25p (40¢) each addl. 5 min. Baggage 15p per piece. Surcharges weekends, after midnight. Tip 10-15%.

Tour operators' courtesy coach meets flights. 10 min.

No. 3/3B bus every 30 min 8:30 am-8:30 pm. Pay driver 20p (35¢). Baggage OK. Route: Winston Churchill Ave., Smith Dorrien Ave., Line Wall Rd., Convent, Southport Gates, Europa Rd., Loreto Convent, Royal Naval Hospital, Lighthouse. Also No. 9 bus every 15 min to Town Centre.

British Car Rental, Avis, Budget, Europacar, Hertz, Niza, Joncab.

P £2 ($3.40)/day.

GLASGOW, Scotland
Glasgow Airport, 9 mi (14 km) W

£9 ($15.65), 15 min.

CityLink Service 500/502. every 30 min, M-Sat 6:25 am-11:55 pm, Sun 7:55 am-10:55 pm, £1.90 ($3.30), 20 min.

Alamo, Avis, Budget, Eurodollar, Europcar, Guy Salmon, Hertz, Kenning.

P ST £4.50 ($7.85), LT £1.95 ($3.30).

GOTHENBURG, Sweden

Landvetter Airport, 15.5 mi (25 km) E

Skr230 ($39) to Central Station, 20 min. No charge first 25 kg of luggage. Any extra, Skr5 (80¢). Tip 10%.

Outside international terminal, every 15 min, 6 am-midnight, Skr50 ($8.50), 30 min.

City Terminal bus from intl terminal 2-3 times hourly 6 am-11:21 pm M-F, 5:45 am-11:15 pm Sat-Sun. Pay Skr25 ($4.10) aboard bus. 30-35 min to Central Station. Good baggage space.

Avis, Budget, Europcar, Hertz, InterRent.

P ST Skr60 ($10.20)/day.

GRAND JUNCTION, Colorado

Walker Field, 6 mi NE

$7.40, 15-20 min.

Avis, Budget, Hertz, National, Payless, Thrifty.

GRAND RAPIDS, Michigan

Kent County Intl Airport, 11 mi SE

$20, 15 min to Amway, downtown. If no taxis at airport, use "Taxi" phone in shelter area outside terminal.

Avis, Budget, Hertz, National.

P ST $10, LT $2.50/day, $15/week.

GRAZ, Austria

Graz Airport, 7.5 mi (12 km) S

S200 ($16), 15 min.

BB bus, S42 ($3.35) to Hotels Weitzer and Wiesler, and station.

Avis, Budget, Europcar, Hertz.

GREEN BAY, Wisconsin
Austin Straubel International Airport, 8.5 mi SW

$9-9.75, extra passengers 50¢ each. 12-15 min.

Avis, Budget, Hertz, National.

P ST $4.50, LT $3.25, weekly $17.50.

GREENSBORO/HIGH POINT/WINSTON-SALEM, North Carolina
Greensboro Regional Airport, 9 mi W of Greensboro, 18 mi E of Winston-Salem, 12 mi NE of High Point

$14, 10-20 min to Greensboro; $27, 30-35 min to Winston-Salem; $14, 20 min to High Point.

Airport Limousine Service. To Greensboro, High Point hourly on the half hour 5:30 am-12:30 am daily. $7.50. To Winston-Salem, hourly on the hour 5 am-midnight. $12-15. AMEX, MC, Visa.

Alamo, Avis, Budget, Dollar, Hertz, National, Thrifty, Triangle.

P ST $4.50, LT $3.50.

GREENVILLE/SPARTANBURG, South Carolina
Greenville/Spartanburg Airport, 15 mi NE of Greenville, 13 mi SW of Spartanburg

$15, 20 min to Greenville; $30, 30 min to Spartanburg. Extra passengers 25¢ each.

Avis, Hertz.

GUADALAJARA, Mexico
Aeropuerto Miguel Hidalgo, 10.5 mi (17 km) S

$1-5, by zone, 10-25 min. Tip 15%.

Chapala-Guadalajara bus picks up outside intl arrivals door, hourly. 25¢, 40 min. Baggage OK.

Alal, Aries, Arrasa, Avis, Dollar, Hertz, Interrenta, National, Odin, Ohara, Quick, Rent-Ford.

P $1.70/day.

GUANGZHOU, China
Baiyun Airport, 7.4 mi (12 km) N

Train station, Dong Fang Hotel, China Hotel (all in center city), RY3 (80¢), 10 min; White Swan Hotel, RY15 ($3.90), 40 min; Garden Hotel, RY10 ($2.60), 20 min. Tip 5-10%.

CAAC bus from airport to city ticket office every 20 min 6 am-11 pm. RY0.50 (13¢), 20 min.

Self-drive cars not available in China.

GUERNSEY, Channel Islands
Guernsey La Villiaze Airport, 4 mi (6.5 km) WSW

£4.50 ($7.85), 15 min to St. Peter Port. Tip 10-15%.

Guernseybus No. C1 & C2 from the airport forecourt every 30 min 8 am-6:35 pm. 75p ($1.30). 20 min. Baggage OK.

Avis, Harlequin Hire.

P £2.50 ($4.35)/day.

GULFPORT/BILOXI, Mississippi
Gulfport Municipal Airport, 4 mi NE of Gulfport, 13 mi W of Biloxi

$9.50, 15 min to Gulfport; $20, 15-20 min to Biloxi.

To New Orleans Intl Airport Coastliner van at 3:30, 5:30, 7:30, 9, 10:30 am; 12:30, 2, 3:30, 5:30 pm. To N.O. Intl, $31 OW, $53 RT, 2½ hr. Info & res (US) 800-647-3957.

Avis, Budget, Dollar, Hertz, National, Thrifty.

HALIFAX, Nova Scotia
Halifax Intl Airport, 25 mi NE

C$27 ($23).

Nova Charters bus meets flights. C$8 ($6.80). Serves all major hotels. 40-50 min ride into city. AMEX, MC, VISA.

Avis, Budget, Hertz, Thrifty, Tilden.

HAMBURG, Germany
Hamburg Airport, 5.5 mi (8.5 km) N

DM30 ($18), 20-25 min into city. Tip: round up fare to nearest mark or DM1.

Airport-City bus every 20 min 6:30 am-10:50 pm from arrival hall. Pay DM8 ($4.80) to driver. Stops at SAS Plaza Hotel, Hauptbahnhof Kirchenallee (main RR station). About 25 min.

No. 110 (No. 606 at night) to Ohlsdorf in northern section of city, where one transfers to city trains and U-Bahn. Bus, DM3.20 ($1.90) including transfer. Good choice for travelers without baggage. 30-45 min.

Autohansa, Auto-Sixt, Avis, Europcar, Hertz, InterRent.

P ST DM30 ($18), LT DM10 ($6).

HAMILTON, Bermuda
Bermuda Air Terminal, 9 mi E

$16, 30 min. To Southampton area, $30. Baggage 25¢ each, maximum $4. Tip 15%. Fares to hotels and guest houses range from $6 to $35 depending on distance. (Bermuda dollar is on a par with US dollar.)

Bermuda Aviation Services coach on demand to hotels. $6.50-$12.50, 15-50 min.

Public bus to Hamilton leaves from S side of main terminal every 15 min 6:45 am-11:45 pm. $2.50, coins only. 45 min. Small baggage OK.

None.

P Free.

HANNOVER, Germany
Langenhagen Airport, 8 mi (12.8 km) N

DM30 ($17.40), 25 min. For tip, round up to nearest mark.

No. 60 direct to air terminal at rear of Hauptbahnhof where there are cabs for easy transfer of baggage. Bus leaves every 30 min 5:20 am-11:10 pm. DM4.70 ($2.80). 20 min trip.

Autohansa, Avis, Europcar, Hertz, InterRent, Sixt (Budget).

HARARE, Zimbabwe
Harare Intl Airport, 10 mi (16 km) SE

Z$25 ($3.75), 20 min. No tip required.

Air Zimbabwe coach departs hourly each direction between airport and city terminal 6 am-11 pm. Z$10 ($1.50), 20 min. Ticket must be purchased in advance at airport.

Avis, Hertz.

HARRISBURG, Pennsylvania
Harrisburg Intl Airport, 11 mi SE

$14-15, 10 min.

Avis, Budget, Dollar, Hertz.

HARTFORD, Connecticut/SPRINGFIELD, Massachusetts
Bradley Intl Airport, 14 mi N of Hartford, 18 mi S of Springfield

Hartford $21 flat rate, 15-20 min. Springfield $26-30, Torrington $40-50, Putnam $60-70, Norwich $60-70, Meriden $30-40, New Haven $60-70, New London $70-80, Bridgeport, $80-90, Stamford, $110-120.

To Hartford Airport Taxi Co. hourly at :15. $8 to Ramada, Sheraton, Holiday Inn. **To Springfield** Peter Pan bus at 7, 9:15, 11:15 am; 1:15, 3:15, 5:15, 6:50, 9 pm; also 10:15 pm ex Sat. $7.95 OW, $15 RT. 30-40 min to bus terminal at Main & Liberty.

Take Employee Shuttle bus from second island opposite baggage claim to Employees' Parking lot. Then "N" bus downtown at 6:18, 6:58, 7:58, 9:22, 10:22, 11:22 am; 12:22, 1:22, 2:22, 3:05, 4:23, 5:27, 6:05 pm. $1.15.

Avis, Budget, Dollar, Hertz, National.

P ST $12, LT $8/$32 per week.

HAVANA, Cuba
Jose Marti Airport, 14 mi S

$11-15 (US currency accepted), 25-30 min. Cabs available all times.

HELSINKI, Finland
Helsinki Vantaa Airport, 12 mi (19 km) N

FIM110 ($26.40), 20-30 min to city center. Surcharge for more than 2 passengers. No tip.

Finnair bus 2-4 times an hour 6:10 am-11:25 pm. Pay FIM17 ($4.10) on bus. 25-40 min to Hotel Inter-Continental, Central Railway Station, other stops on request.

No. 614 to Central Bus Station, No. 615 to Central Railway Station. Several stops along way. Baggage space but less than on limo. Pay FIM13 ($3.10) aboard. Departures every 20 min from 5:25 am weekdays, 6:10 weekends, to 11:05 pm. Not A/C. 30-40 min ride.

Avis, Budget, Hertz, Inter-Rent/Eurocar.

P ST FIM10 ($2.40)/hour. LT FIM40 ($9.60)/day covered, FIM25 ($6) uncovered.

HERMOSILLO, Mexico

Hermosillo Airport, 3.7 mi (6 km) W

$7, 10-15 min.

Avis, Budget, Hertz.

HILO, Hawaii

Hilo Intl Airport, 3 mi E

$6, 8 min. Baggage, 30¢ each. Bicycles, surfboards extra.

Alamo, Avis, Budget, Dollar, Hertz, National, Travelers.

P $6/day.

HILTON HEAD ISLAND, South Carolina

Hilton Head Airport, 10 mi NE of Sea Pines Plantation

$12 flat, 20 min. If no cab at airport, call Yellow: 686-6666.

American Intl, Avis, Budget, Hertz, National.

HOBART, Tasmania

Hobart Airport, 12.5 mi (20 km) E

A$18 ($14), 15-20 min. To Sandy Bay, A$18 ($14), 20-25 min; to Northern Suburbs, A$20 ($15.40) 25 min; to Howrah, A$14.50 ($11.20), 20 min. Tip not necessary.

Red-Line meets flights, departs from outside terminal. A$3.50 ($2.20), 20 min to city center, casino.

Avis, Budget, Hertz, Thrifty.

P A40¢ (30¢)/day.

HONG KONG

Hong Kong Intl Airport, 3 mi (4.8 km) NE of Tsim Sha Tsui, Kowloon; 6 mi (9.7 km) N of central Hong Kong Island

To Tsim Sha Tsui, Kowloon, HK$36 ($4.70), 15-20 min. To central Hong Kong Island including Cross Harbour Tunnel fee, HK$72 ($9.30), 35-45 min.

To Causeway Bay, Hong Kong Island including tunnel fee, HK$65 ($8.30) 25-35 min. Pay fare on meter in Hong Kong currency. Tip: Round up to next HK$.

A1 Airbus Kowloon hotels/guest houses: Ambassador, Empress, Grand, Holiday Inn Golden Mile, Omni the Hongkong, Hyatt Regency, Imperial, International, Kowloon, New Astor, New World, Park, Peninsula, Ramada Hotel Kowloon, Sheraton, Windsor, YMCA, Chungking House, Guangdon, Kimberley, Kowloon Shangri-La, Miramar, and Regent. HK$8 ($1), 30 min ride. Departures every 15 min 7 am-midnight. **A2 Airbus** Central/Wanchai hotels: Furama Kempinski, Mandarin Oriental, Victoria, Grand Hyatt, Harbour View Intl House, New World Harbour View, Evergreen Plaza, Harbour, Hongkong Hilton, Luk Kwok, New Harbour, and Wharney. HK12 ($1.55). 45-min ride. Departures every 15-20 min 6:50 am to midnight. **A3 Airbus** Causeway Bay hotels: Excelsior, Lee Gardens, Park Lane. HK $12 ($1.55). 45-min ride. Departures every 15-20 min 6:55 am to midnight.

P HK$12 ($1.55)/hr first 2 hrs; HK$24 ($3.10) per hour each addl.

HONOLULU, Hawaii

Honolulu Intl Airport, 4.5 mi NW of financial district, 8 mi NW of Waikiki

$10-12 to business district, 10-15 min; $16-21 to Waikiki, 25-30 min; $21-23 to Diamond Head, 30-35 min.

Airport Motorcoach every 20 min to Waikiki. $5. Reservation required for return to airport: 926-4747. Other vans: Waikiki Express, 942-2177. Airport Express, 949-5249. Celebrity Express 923-5700. Airport Super Shuttle, 261-7880.

No. 19 or No. 20 bus from ramp every 20-30 min 6:28 am-11:30 pm. 60¢ exact. Baggage must be small enough to place on lap or under seat. 20-25 min to business district, 40-50 min to Waikiki. With free transfer, it costs only 60¢ to go anywhere on the island.

Alamo, Avis, Budget, Dollar, Hertz, Honolulu, National, Sunshine, Thrifty, Tradewinds, Tropical.

P $8.70/day.

HOUSTON, Texas

William P. Hobby Airport, 9.5 mi SE

$15-17, 30 min. To Galleria, $25-27, 30 min. To Houston Intercontinental Airport, $35-40, 60-75 min in heavy traffic. United Taxi accepts AMEX, Diners, Discover, Visa, MC.

Hobby Airport Limousine to Hyatt Regency, downtown; South Main; Post Oak, near Galleria; Greenway Plaza. $5. Every 30-60 min 7:30 am-

12:30 am. **To Galveston** Galveston Limousine hourly 8 am-Midnight. $15 to terminals, hotels, Medical Center; $3 addl to private residences. Serves other cities en route: see entry for Houston Intercontinental Airport, below.

No. 50 Heights or Downtown bus, 85¢ exact. About every 20 min 4:17 am-12:12 am M-F. Every 30-35 min Sat-Sun-Hol. 50 min downtown.

Alamo, Avis, Budget, Dollar, General, Hertz, National, Thrifty, and others.

P ST $20/day, LT $7/day.

HOUSTON, Texas

Houston Intercontinental Airport, 22 mi N

$28.75 downtown, 30-50 min; NASA $46. $1.25 airport charge added to fare. Ride sharing available.

Airport Express every 30 min 6:50 am-12:50 am to Hyatt Regency downtown, Galleria, Post Oak terminal/Greenway Plaza. $9.35. To Medical Center, South Main & Astrovillage: every 45 min 7:35 am-12:50 am. $9.35. **To Town & Country** Town & Country Airport Shuttle hourly at :05, 6:05 am-11:05 pm. $15. Serves Adam's Mark, Westchase Hilton, Holiday Inn Houston West, Hilton West, Hyatt Regency, Sheraton Town & Country. **To Galveston** Galveston Limousine every 2 hrs 8 am-10 pm & 11 pm. To Galveston hotels, $18; $3 addl for residential drop off & pickup. Also serves Texas City, LaMarque, Hitchcock, Santa Fe, Dickinson, League City, NASA/Webster, Pasadena, Alvin. Info: 713-223-2256.

Alamo, Avis, Budget, Dollar, General, Hertz, National, Payless, Snappy, Thrifty.

P ST $8/$7/$3.

HUNTSVILLE, Alabama

Huntsville International Airport, 11 mi SW of Huntsville

$14.50 to Huntsville, 15-20 min. Decatur, $35.

Airport Limousine Service 6:30 am to last flight. $11-12 per person to Huntsville addresses, $15-20 to Decatur.

Avis, Budget, Dollar, Hertz, National.

HYDERABAD, India

Hyderabad Airport, 12 mi (19 km) NW

Rs20 ($1.15), 25 min, to city center, RR station. To Secunderabad RR station, Rs15 (85¢), 20 min. To Hyderabad Bus Depot, Rs25 ($1.40), 35 min.

Baggage Rs5 (30¢) each. Tip Rs2 (10¢).

Rental cars not available in India.

INDIANAPOLIS, Indiana

Indianapolis Intl Airport, 8 mi SW

$14, 15 min downtown. $27 East Side; $35 Ft. Benjamin Harrison.

Shuttlexpress van 7 am-11 pm. $7 to downtown hotels. **To Lafayette, Purdue** Lafayette Limo every 2 hrs 6:30 am-10:30 pm. Info: 317-497-3828. **To Noblesville, Anderson, Ball State, Muncie** American Bus Lines 6 am-11:30 pm. Info: 800-228-0814 (Indiana only).

Service every 30-60 min 5:03 am-12:18 pm M-F. Sat hourly 7:33 am-7:15 pm. Sun-Hol hourly 7:30 am-6:30 pm. $1.10.

Alamo, Avis, Budget, Dollar, Hertz, National.

P ST $9, LT $3.50. Shuttle.

INVERCARGILL, New Zealand

Invercargill Airport, 2 mi (3.2 km) W

NZ$8-9 ($4.65-5.20), 5 min. No tip. To Bluff, NZ$25 ($14.50), 30 min. Cabs do not wait at airport. Use free phone to call Blue Star Taxi.

H&H Travel Line meets flights. NZ$1.50 (90¢), 10 min.

Avis, Bluff, Budget, Hertz.

ISTANBUL, Turkey

Yesilkoy (Ataturk) Airport, 15 mi (24 km) W

T£45,000 ($9.85), 30-50 min to Taksim Sq; T£35,000 ($7.65), 30 min to Aksaray. Small orange cabs, not A/C. Baggage T£2000 (50¢). Tip optional.

HAVAS bus every 30 min 6 am-10 am; then hourly to 2 pm; then every 30 min to 8 pm; then hourly to 11 pm. To Sishane, Aksaray 30-50 min. Pay fare T£12,000 ($2.65) aboard bus. Comfortable.

No. 96 bus every 55 min 7 am-9 pm from outside terminal to Aksaray Sq, making several stops en route. Buy ticket for T£1500 (33¢) at kiosk. Ride can take from 30 min to an hour depending on traffic. Not A/C but otherwise comfortable ride for traveler with light baggage.

Avis, Budget, Camel, Esin, Europcar, Hertz, Inter-Limousine, InterRent, Kontuar, Kayhan, Ledz.

P T£15,000 ($3.30)/day.

IZMIR, Turkey
Adnan Menderes Airport, 10 mi (16 km) S

T20,000 ($4.40) flat rate for up to 3 persons. 20 min to city. Small cabs—Fiats, Renaults.

Airport Bus operated by Turkish Airlines serves passengers on all carriers. T2000 (45¢), 25 min to Efes Hotel.

American Intl, Avis, Hertz, InterRent.

JACKSON, Mississippi
Jackson Municipal Airport, 10 mi E

$12, $1.50 each addl passenger. 20 min.

Avis, Budget, Hertz, National.

P ST $6, LT $4.

JACKSONVILLE, Florida
Jacksonville Intl Airport, 17 mi N

$19 plus tolls, addl riders $1 each. 20-60 min.

Avis, Budget, Dollar, Hertz, National.

P Garage, $6.50; uncovered, $4.50/day.

JAKARTA, Indonesia
Soekarno-Hatta Intl Airport, 16 mi (26 km) W

Rp17,000-20,000 ($10-12), 50-60 min to Borobudur, Hilton, Mandarin. Rp400 (25¢) per bag. A larger tip is expected if more than one person rides in cab but Rp1700 ($1) should be OK.

Avis, National.

JEDDAH, Saudi Arabia
King Abdulaziz Intl Airport, 20 mi (32 km) S

R50 ($13), 25-35 min. Tip not customary.

Abu Ziab, Al-Ebery, Al-Premy, Arabian, Arabian Halar, Avis, Hanco, Jama, Key, Shary, Universal.

P R24 ($6.40)/day.

JERSEY, Channel Islands
Jersey Airport, 5.5 mi (9 km) WNW of St. Helier

£6 ($10) tip included, 15 min.

Jersey Motor Transport every 20 min 7 am-10 pm. Pay driver 80p ($1.35). Stops at Mermaid Hotel, Quennevais, St. Aubin, Grand Hotel, and Weighbridge Bus Station, St. Helier. 30 min to St. Helier.

Europcar, Hertz.

P First hour free. Then 20p/hour, £4.60 ($7.75) first day, £3.60 ($6) subsequent days.

JOHANNESBURG, South Africa
Jan Smuts Airport, 14.5 mi (23 km) NE

Rn46 ($16), 30 min. Tip optional.

To Johannesburg Rn13.50 ($4.75), Pretoria Rn14.50 ($5.10). Pay aboard. Hourly on the half-hour, 5 am-11 pm.

Avis, Budget, Imperial.

P ST Rn10 ($3.50), LT Rn5 ($1.75)/day.

JUNEAU, Alaska
Juneau Intl Airport, 10 mi N

$13.50 to downtown hotels, 15 min. To Mendenhall Glacier Visitor Center, 5 min; to Auke Bay Ferry Terminal, 10 min. Addl riders 50¢ each.

Service after each flight. $6, 15 min. To Baranof, Breakwater, Westmark, Bergmann, Alaskan Hotel, ferry terminals.

Capital Transit System departs approx hourly 7:30 am-11:30 pm M-Sat. Pay driver $1. 20 min. Limited baggage space.

Avis, Budget, Hertz, National, Rent-A-Wreck.

P ST $9, LT $4.50.

KAHULUI, Maui, Hawaii
Kahului Airport, 2 mi NE

$7-8, 10 min to Kahului. Lahaina-Kaanapali $35-40, 40-60 min; Kihei $25, 20-30 min; Wailea (Inter Continental, Stouffer, Four Seasons) $26-28, 30-40 min; Makena Resort area (Maui Prince) $27-30; Wailuku $10; Kapalua $50. For each surfboard or bicycle $3. Yellow Cab, 877-7000.

Trans Hawaiian to Lahaina, Kaanapali hotels hourly 7 am-5 pm. $13. Reservation required day prior.

Alamo, Andres, Avis, Budget, Hertz, Maui Discount, Sunshine, Thrifty.

P $6/day.

KAILUA-KONA, Hawaii
Ke-Ahole Airport, 8 mi NW

To hotels: King Kamehameha $15-16; Kona Hilton $16-17; Kona Surf $26; Kona Village $16; Mauna Kea Beach $45; Royal Waikoloa $32; Hyatt Waikoloa $33.

Alamo, Avis, Budget, Dollar, Hertz, National, Roberts, Tropical.

KALAMAZOO, Michigan
Kalamazoo County Airport, 4 mi SE

Downtown, Upjohn, $6.40, 10 min. To Battle Creek, $30 flat rate, 30-40 min. Extra passengers 50¢ each.

Avis, Budget, Hertz, National.

P ST $12, LT $2.50. Parking lot closed 11:30 pm-5:30 am and 11 pm Sat-6 am Sun.

KANSAS CITY, Missouri
Kansas City Intl Airport, 18 mi NW

$21-29 downtown, 25-40 min. $22-30 Crown Center, Country Club Plaza. $26-35 Overland Park, $33-45 Cab fares are deregulated and vary from company to company. Zone fares and meter rate apply and the passenger pays whichever is less. Best to confirm fare in advance with cab driver.

KCI Express every 45 min 5:45 am-11 pm. 25-40 min to downtown hotels. $9 OW, $17 RT. To Crown Center $10, Country Club Plaza, $11. To Johnson County hotels, 7 am-10:15 pm hourly. $14 OW. MC, VISA accepted. Leaves from Terminal C, Gate 63 (take airport shuttle if that's too long a walk).

No. 29 KCI airport bus from Terminal A at 6:54, 7:16, 7:47 am; 2:58, 3:56, 4:52 pm. Also picks up at terminals B, C. 85¢. 55 min to downtown KC, 65 min to Plaza. No special baggage space. To catch this bus you must go down ramp at end of terminal to outer circle roadway and stand next to parking lot. No weekend, holiday service.

Alamo, Avis, Budget, Dollar, Hertz, National, Thrifty.

P ST $10/day, LT $4-8.

KAUAI, Hawaii
Lihue Airport, 2 mi E of Lihue

To Coco Palms Resort, $15; Sheraton Princeville, $55; Kapaa, $18; Lihue Convention Center, $5. Cabs are expensive on Kauai. Tourists staying more than a couple of days will do better renting a car, locals advise.

Alamo, Avis, Budget, Dollar, Hertz, National, Roberts, Tropical.

KEY WEST, Florida
Key West Intl Airport, 1 mi SE

$4 flat rate, 10 min to Duval St.

Avis, Dollar.

KINGSTON, Jamaica
Norman Manley Intl Airport, 11 mi SE

J$70 ($13.35), 40 min.

American Intl, Avis, Budget, Dollar, Hertz, National.

KNOXVILLE, Tennessee
Knoxville Airport, 12 mi S

$14 for 1-2 passengers, $7 each 3 or more. 25-30 min.

Airport Limousine leaves 10 min after each flight arrives. $12 to downtown hotels.

Avis, Budget, Dollar, Hertz, National, Thrifty.

KRALENDIJK, Bonaire
Flamingo Airport, 2 mi (3.5 km) S

NAfl9 ($5), 5 min. To Flamingo Beach Hotel, NAfl7 ($4); Bonaire Beach Hotel;NAfl14 ($8); Habitat, NAfl15 ($8.50); Entrejol, NAfl12 ($7); Oil Terminal, NAfl36 ($20); Playa Grandi, NAfl36 ($20); Punta Blancu, NAf17.5 ($10); Rode Pan, NAfl27 ($15); Zuidkust (AISCO), NAfl23 ($13); Rincon, NAf23 ($13); Sand Dollar Beach Club, NAf14.40 ($8). Surcharges: 8 pm-midnight, 25%; midnight-6 am, 50%. More than 4 passengers, 20% each addl. Tip 10%.

AB Carrental, Avis, Boncar/Budget, Evertsz, Sunray, Total Carrental.

P Free.

KUALA LUMPUR, Malaysia
Subang Intl Airport, 14 mi (22.5 km) SW

RGT22 ($8.15), 30-40 min. Check posted rates. Buy coupon at booth, give to driver. Tip optional. 50% surcharge midnight-8 am.

Comfortable A/C coach, RGT5-10 ($1.85-3.70), 30 min.

Avis, Europcar, Hertz, Sintat.

LA PAZ, Mexico
General Marquez de Leon Airport, 7 mi (11.3 km) N

$7, 15 min.

Avis, Budget, Hertz.

LAHORE, Pakistan
Lahore Airport, 2 mi (3.2 km) NW

Hilton Intl, Pearl Continental provide free airport service.

Avis.

LAKE TAHOE, California
Lake Tahoe Airport, 5 mi SW

To Casino Core, $13.50, 15-20 min.

RTS coach meets flights. $5.00, 15-20 min to Harrah's, Harvey's, High Sierra, Caesar's, etc. Purchase tickets in advance in terminal.

To Reno Intl Airport from casinos $12.50, 90 min.

Avis, Budget, Hertz.

P $3/day; $50/month.

LANSING, Michigan
Capital City Airport, 4 mi NW

$8-9 downtown, 10 min.

Yellow Cab operates van downtown. $4 per person. Meets most flights—use white courtesy phone to call.

Avis, Budget, Hertz, National.

P ST $7, LT $3/day, $15/week.

LARNACA, Cyprus
Larnaca Intl Airport, 4 mi (6.4 km) S of Larnaca, 33 mi (53 km) SE of Nicosia

To center of town Cy£1.50 ($2.50). Shared taxis serve entire island from Larnaca. Expect to pay about Cy£1.25 ($2.10) per person for a 50-mile ride—to Limassol, for example.

Avis, Budget, Eurodollar, Glamico, Hertz, A. Petsas & Sons.

LAS VEGAS, Nevada
McCarran Intl Airport, 5 mi S

$11 to The Strip, $15 downtown.

Gray Line, Bell Trans, Lucky. To Strip, $3.25-$3.75. To Downtown, $4.25-$4.75. Departures every 20 min, 5 am-2 am. Stops at all major hotels. Return schedule available in hotel lobby.

Allstate, Avis, Dollar, Hertz, National, Sav-mor, and others.

P ST $6-$72 max/month, LT $4-$48 max/month.

LEIPZIG, Germany
Leipzig Airport, 7.5 mi (12 km) NW

DM30 ($17.40), 20 min to Hotel Merkur, center city.

Avis.

LENINGRAD, USSR See ST. PETERSBURG, Russia

LEXINGTON/FRANKFORT, Kentucky
Blue Grass Field, 6 mi W of Lexington

$10, 15 min to Lexington; $22 flat, 40 min to Frankfort.

Avis, Budget, Dollar, Hertz, National, Snappy, Thrifty.

P ST $6.50, LT $4.25/day.

LIMA, Peru
Jorge Chavez Intl Airport, 5 mi (8 km) W

About $11 downtown, no extras, no tip. Agree on price before getting in cab. 30-45 min. Colectivos—shared-ride cabs—charge about half of regular cab price, operate from 6 am-6 pm. Stop at La Colmena 73, midway between Gran Bolivar and Crillon hotels.

Trans Hotel bus every 20 min. $5. Stops at hotels and other places on request. Also inquire about hotel vans.

Airport Bus at Enatru station outside airport. Hourly, $1.50, baggage space. Terminus at Ave. La Colmena. Long ride—90+ min.

Avis, Budget, National, Hertz.

LINCOLN, Nebraska
Lincoln Municipal Airport, 5 mi NW

$7.50, 10-15 min.

Avis, Budget, Dollar, Hertz, National.

LISBON, Portugal
Lisboa (Portela) Airport, 4.5 mi (7 km) N

Esc650 ($5.45), 15-20 min to Ritz. Taxi rank at left outside arrival hall. For several bags driver can charge 50% more than meter. Tip 10%.

Green Line bus every 15 min 7:30 am-9:20 pm M-Sat, every 20 min Sun. Pay driver Esc275 ($1.80). 35-40 min to Santa Apolonia RR station by river in old Alfama section. Little baggage room. Bus connects with subway at Entrecampos.

Avis, Budget, Europcar, Guerin, Hertz/Renorte, InterRent/Global Rent, Travelcar, Turincar.

P ST Esc2546 ($16.90).

LITTLE ROCK, Arkansas
Little Rock Regional Airport (Adams Field), 3.5 mi SE

$7.50, each addl rider $1. 10-15 min.

To Hot Springs Collins Airport Service about every 2 hrs 8:30 am-11 pm. $18 per person.

Avis, Budget, Dollar, Hertz, National.

LONDON, England
Gatwick Airport, 28 mi (45 km) S

£43 ($74.80) to central London. Tip 10-15%. 60-90 min trip depending on time of day. **To Heathrow Airport:** £42 ($73), 60-90 min.

No. 777 bus hourly on the hour 7 am-10 pm. Picks up at both North and South terminals. £6 ($10.45) OW, £8 ($13.90) RT, 70 min to Victoria Coach Station, half mile from Victoria Rail Station.

Gatwick Express every 15 min 6:20 am-10:50 pm; then at 11:20, 11:50 pm, 12:10 am; then hourly at :05 to 6:05 am. £7 ($12.20). Wide doors, automatic inside doors, ample baggage space. 30 min express ride to Victoria Station. No service Christmas Day.

To Heathrow Airport Speedlink every 15-30 min 6 am-10 pm. Fare £13 ($22.60) adult, £7 ($12.20) child. 60-min ride to Heathrow, very comfortable. No service Christmas Day.

Avis, Budget, Europcar, Hertz.

P ST £8.60 ($14.95), LT £3.65 ($6.35).

LONDON, England

Heathrow Airport, 15 mi (25 km) W

To Central London £27 ($47), 30-50 min. Addl passengers 30p (50¢) each, bags 20p (20¢) each. Evening and late-night surcharges of 40-60p (70¢-$1); also, 8 pm Dec 24 to 6 am Dec 27, and 8 pm Dec 31 to 6 am Jan 1, £2 ($3.50) extra. Tip porter 40p per bag; tip driver 10-15%. Cab sharing available in cabs so marked. Fare each passenger pays as percentage of full fare: two, 65%; three, 55%; four, 45%; five, 40%.

Flightline 767 bus from each terminal and Central Bus Station to Victoria Coach Station every 30 min 5 am-5:15 pm, hourly to 9:15 pm. Fare £5 ($8.40) OW, RT same day £6 ($10), RT later £7 ($11.80). 65 min. Driver loads baggage, very comfortable ride, onboard WC.

Airbus A1 to Victoria Station every 30 min 6 am-9 pm. A2 to Euston Station every 30 min 6 am-9 pm. 50-60 min ride. Fare £5 ($8.70) OW, £8 ($13.90) RT. Ample baggage space, comfortable ride. These magnificent red double-deckers are a sightseer's delight and fast. They stop at major hotel areas en route.

Heathrow terminals are stations on Piccadilly Line. Moving walkways to station—baggage carts can be taken to entrance. Buy ticket at counter or from machine, £1.90 ($3.20) to Piccadilly. Comfortable for travelers with small baggage. Trains every 3 min peak times, every 7 min off-peak & Sun, from 5:08 am-11:49 pm Mon-Sat, 6:01 am-10:27 pm Sun. No service Christmas Day. 47-min average to Piccadilly, 55 min to King's Cross, 61 min to Liverpool St. Note there may be stairs to climb at some exits.

To Gatwick Airport Speedlink 747 from Terminal 4 and Central Bus Station hourly 6:15 am-10:30 pm. Fare £13 ($22.60). 60 min to Gatwick terminal. No service Christmas day.

Alamo, Avis, British, Budget, Eurodollar, Europcar, Hertz, Kenning, and others.

P ST £1.60 ($2.80)/hr, over 7 days £12 ($21.90)/day; LT £6.90 ($12) per day through 7 days, £4.90 ($8.55) per day after 7 days.

Bus connections elsewhere Many bus routes connect Heathrow with the surrounding countryside. Inquire at Central Bus Station.

LONDON, ENGLAND

London City Airport, 6 mi E

£8 ($13.50), 25 min to the City. Advise airplane cabin crew during flight and a taxi can be ordered in advance of landing.

LONDON, England

Stansted Airport, 37 mi (59 km) NE of London, 4 mi (6.4 km) NE of Bishop's Stortford

Aircars Taxi Service has a counter in arrivals hall. £4 ($6.70), 10-15 min to British Rail Station in Bishop's Stortford, where there is service to Liverpool Street Station, London, a 45-50 min ride, for £5.30 ($8.90) standard class. Service about twice an hour, more at peak times. Hourly Sundays. Stansted Airport Cars to London, £35 ($61), 45-60 min.

Rail Air Link Buses operate between airport and Bishop's Stortford Station. Service about every 30 min. £1.05 ($1.80). 15-min ride.

To Heathrow, Gatwick Premier Travel coach every 2 hrs 4:20 am-6:20 pm. Speedlink 8:20, 10:50 am, 12:50, 2:50, 4:50, 8:05 pm. £12 ($21). 80 min to Heathrow, 1 hr 40 min to Gatwick.

Avis, Budget, Europcar, Hertz.

P ST £3.15 ($5.50). LT £2.15 ($3.75).

LONG BEACH, California

Long Beach Municipal Airport, 10 mi N of Long Beach, 24 mi SE of Los Angeles

To Hyatt Regency, $15; Queen Mary area of Long Beach, $17, 20 min. To cruise port, $25-27. To LAX, $35, 40-50 min; to downtown LA, $40-45, 40-70 min.

To Long Beach Harbor (San Pedro) SuperShuttle $20 first passenger, $7 each addl. **To Disneyland** SuperShuttle $33 first psgr, $7 each addl. **Anaheim & Orange County destinations** SuperShuttle info & res: 800-554-6458 or 213-338-1111. AMEX, Visa, MC accepted.

To LA No. 457 RTD Express Bus M-F am only from Airport Park-Ride lot at 6:00, 6:13, 6:31, 6:49, 7:07, 7:25, 7:45, 8:05 am. $2.70, bills OK. Hour ride to 5th & Flower, downtown. Room for small baggage. Catch express return from 6th & Flower pm. Ask driver for schedule. **To Long Beach** No. 111 Broadway/Lakewood bus every 30-60 min 5:46 am-1:02 am M-F; hourly 6:12 am-11:25 pm Sat-Sun-Hol. 60¢.

To LAX SuperShuttle $23 first psgr, $7 each addl. Reserve a day in advance: 800-554-6458 or 213-338-1111. AMEX, MC, VISA. **To Orange County Airport** $32.

Avis, Budget, Dollar, Hertz, National.

P ST $12.60; LT $9.45/day.

LOS ANGELES, California

Los Angeles Intl Airport, 15 mi SW

INFO, SCHEDULES Inquire at "Ground Transport" booth outside each terminal. Airport info by phone: 213-646-5252.

Downtown LA $26.50, 30-35 min. Anaheim/Disneyland, $85. Long Beach Airport/downtown, $46-50. Century City $24-26. Fares include $2.50 airport tax. Load at yellow "TAXI" sign.

Door-to-door Vans. 37 or more on-demand operators serve hotels, offices, and residences throughout Los Angeles and Orange, San Bernardino and Riverside counties. Several operators accept major credit cards. Fares vary greatly. Inquire at "Ground Transport" information booth for fares and info. Wait at blue "Van Stop" sign on island outside lower level. With reservation or operator name, van will be dispatched immediately by island agent. With destination only, dispatch will be notified. If van to your destination dispatched in last 5 mins, agent will flag them down. If not, van will be dispatched from holding lot. Ride-share vans offer convenience compared to scheduled bus, cost saving compared to taxi in many cases. But trade-off is time waiting for van to arrive and further delay if driver circles for other fares or drops you off last. Inter-airport connections made by some operators. Inquire at info booth.

Scheduled Coaches. 7 operators. Areas served: Anaheim (to specific hotels), $13-$14; John Wayne Airport (via 3 Disney hotels, $18; Santa Barbara, Goleta, Isla Vista & Carpenteria, $33; Bakersfield, $25; Lancaster, Palmdale, $18; Newhall, $12; Pasadena (to specific hotels), $12; Ventura

County, $20; and Van Nuys (FlyAway Van Nuys Terminal), $3.50. Inquire at "Ground Transport" info booth for schedules.

All public buses to downtown and elsewhere board at LAX Transit Center just outside airport. Free shuttle bus "C" at "LAX Shuttle" sign every 10 mins 24 hrs a day. 7-min ride to transfer point. The center is served by RTD Bus, Culver City Bus, Torrace Transit and Santa Monica City Bus. Downtown LA No. 439 freeway express Mon-Fri from Bay 11 between 5:11 am & 10:59 pm. Operates roughly every 30 mins except late at night. Fare $1.50. Line 42 operates between 5:34 am and 11:20 pm at least every hour & more frequently at busy times of day. Fare $1.10. 45-50 mins to downtown. Other destinations served: West Hollywood, Palos Verdes Peninsula, Long Beach, South Bay, Whittier, Lynwood, Lakeview Terrace, Westwood, Norwalk, Marina del Rey, Torrance, Venice & Santa Monica. Info: (213) 626-4455.

Interairport Connections To Burbank, Long Beach, John Wayne & Ontario airports. SuperShuttle: call 417-8988 from LAX; elsewhere 213-338-1111.

Avis, Budget, Dollar, Hertz, National.

P ST $16; LT $7-5, first 2 hrs free. Free shuttle buses "B" and "C" to long-term lots every 10 min 24 hrs a day. Parking info: 213-646-2911.

SCHEDULED SERVICE TO OTHER POINTS

Van Nuys & San Fernando Valley FlyAway bus from each terminal every 30 min 5:30 am-midnight. Then 12:15, 12:30, 1:15, 2:15, 2:45, 3:30, 4:15, 4:45 am. $4.50 OW, $8 RT. Service to Van Nuys Airport Bus Terminal (7610 Woodley, corner of Saticoy). Parking at Van Nuys available for $1/day up to 30 days max. Info: 818-994-5554. **Lancaster, Palmdale, Newhall** Antelope Valley Airport Express. 8 departures between 7 am and 10:30 pm. Info & res: 805-945-2LAX. **Ventura, Oxnard, Camarillo, Thousand Oaks, Westlake Village, Woodland Hills** Great American Stage Line every 75-90 min 7 am-11 pm. Info: 805-375-1361. **Pasadena** Airportcoach every 60-90 min 6:40 am-9:30 pm. $10 OW, $18 RT. Res: 714-491-3500. **Santa Barbara** Santa Barbara Airbus every 1½ - 2 hrs 7:30 am-11 pm. $26 OW, $48 RT. 2½ hr trip. Also serves Carpinteria, Montecito, Goleta, Isla Vista. Res: 800-423-1618 (CA) 800-733-6354 (US) **Bakersfield** Airport Bus of Bakersfield at 7:30, 10:30 am; 2:30, 6:30, 10:30 pm. 2½-hr trip. $25 OW $40 RT. **Disneyland via Buena Park** Airport Cruiser every 45-60 min 6 am-11:20 pm. $9 OW, $12 RT. 714-761-3345.

L.A. Helicopter connections to Burbank, City of Commerce, City of Industry, Long Beach Airport. Info & res: 213-642-6600.

LOUISVILLE, Kentucky

Standiford Field, 5 mi S

$13 flat, 15-20 min.

Airport Limousine hourly 9:10 am-6:10 pm, $4.50. Meets flights other times. No Sat service. Sun hourly 3:30-9:30 pm. Stops at Galt House, Seelbach, Hyatt.

No. 2 Second St. bus at 6, 7:08, 8:12, 9:55, 11:31 am; 1:07, 2:11, 2:58, 3:36, 4:50, 5:24, 7:43, 8:44 pm M-F. 35 min to 1st & Market downtown. 60¢.

Agency, Alamo, Avis, Budget, Dollar, Hertz, National, Thrifty.

P ST $12, LT $4.

LUBBOCK, Texas

Lubbock Intl Airport, 6 mi N

$9.50, $2 each addl rider. 12 min.

Advantage, Avis, Budget, Hertz, National.

LUTON, England

Luton Intl Airport, 3 mi (4.8 km) E of Luton, 31 mi (50 km) NW of London

Cabco £3.20 ($5.35), metered. Circle Cars £3 ($) flat fare. 5-15 min. Tipping not customary.

Avis, Inter-City, Swan National.

P ST £4.90 ($8.20), LT £2 ($3.35).

LUXEMBOURG City

Findel Airport, 3.75 mi (6 km) E

LF500-600 ($14-17), 15-20 min to Central RR Station. For more than one bag LF20 (60¢) each extra.

Luxair bus every 30 min 6:05 am-10:40 pm. 15-20 min to Air Terminus Luxair near Central Station. Pay LF120 ($3.40) aboard. Not A/C.

Icelandic passengers May use free motorcoach transfer to principal points in Belgium, Holland, W. Germany. Info, reservations: 800-223-5500. **Luxavia passengers from Johannesburg** Similar service into W. Germany.

Luxembourg Ville every 30 min from main terminal. Stops at Youth Hostel, City Center, Central Station. Pay LF25 (70¢) aboard. No special place for baggage but comfortable ride of 15-20 min.

Avis, Budget, Continental/Lux, Europcar, Hertz, InterRent.

P ST LF200 ($5.70), LT LF100 ($2.80).

LYON, France

Lyon/Satolas Airport, 15.5 mi (25 km) E of Lyon; 65 mi (105 km) NW of Grenoble; 75 mi (122 km) SW of Geneva, Switzerland

F140 ($25.20), 20 min; night fare, F200 ($36). To Eurexpo F100 ($18), 10 min. To Grenoble, F550 ($99), 60 min.

Satobus coach departs every 20-30 min 6 am-11 pm. Pay driver F40 ($7.20). 45-min ride stopping at Gare de la Part-Dieu, Perrache Station, Jean Mace, Nouvelles Galeries, Mairie du 8e. To Grenoble, F110 ($19.80). 60 min.

Coach service also available to Grenoble, Annecy, Aix-les-Bains, Chambery, Valence. Inquire at airport information desk.

Avis, Budget, Citer, Europcar, Eurorent, Hertz.

P ST F26 ($4.50), LT F23 ($4).

MADISON, Wisconsin

Dane County Regional Airport, 5 mi NE

$8.50, 12 min to Capitol; $9-11, 12-15 min to University area.

Airport Limousine meets flights: $6 to Capitol, $7 to University.

Budget, Dollar, Hertz, National.

P ST $6, LT $3.

MADRID, Spain

Barajas Airport, 7.5 mi (12 km) NE

Pta1600-1800 ($15-17) to Plaza Colon, 30 min. Extras: Pta50 (50¢) nights, Hol, Pta50 (50¢) each bag. Make certain meter is at zero upon entering cab. Tip 10%.

Airport Bus outside arrival terminal. Pta210 ($2), pay upon boarding. Every 30 min 4:45 am-1:51 am. Comfortable ride, plenty of baggage space. A/C. Stops at Avenida America, Francisco Silvela, Maria de Molina, Velazquez, Serrano, Ortega y Gasset, Plaza Colon.

Avis, Europcar, Hertz, Atesa.

P Pta405 ($3.85)/day.

MALAGA, Spain

Malaga Airport, 7 mi (11 km) SW

Pta1116 ($10.60) to city center, 20 min; Torremolinos, Pta846 ($8), 15 min; Fuengirola, Pta2376 ($22.55), 30 min; Marbella, Pta4986 ($47.35).

No 19, every 30 min 6:30 am-midnight. Boards in front of Gate B, Intl Arrivals Terminal. Fare Pta100 (95¢).

Boards on elevated platform through parking lot and across overpass (up stairs), about 1 mi from terminal. Departures every 20 min 6:30 am-midnight. Pta90 (85¢). Westbound trains also board at this station for Fuengirola and stops in between.

Avis, Benelux, Europcar, Hertz, Lido.

P Pta860 ($8.15)/day.

MANCHESTER, England
Manchester Airport, 10 mi (16 km) S

£9-12 ($15-20) to central Manchester, 20-35 min. For journeys outside city, determine fare in advance with driver.

Airport Bus Stn adjacent to Intl Arrivals hall. Stops at National Coach Stn, Piccadilly Bus Stn, G-Mex Exhibition Centre, Victoria Rail Stn. Every 30 min 6:55 am-9:55 pm. Pay £1 ($1.75) to driver. 40-45 min.

Avis, Europcar, Hertz.

P ST £6 ($10), LT £2.60 ($4.50).

MANCHESTER, New Hampshire
Manchester Airport, 5 mi SE

$9, 50¢ each addl rider. 15 min.

Avis, Budget, Hertz, National.

P ST $12, LT $4/day.

MANILA, Philippines
Ninoy Aquino Intl Airport, 5 mi (8 km) S

R&E, MIATDA taxis, yellow colored, A/C, metered. PP100 ($4.40), 15 min to city center. Non A/C charge less. Pay in pesos, tip optional. Travellers are cautioned not to use any vehicle except an official, metered cab.

Metro Manila Transit Love Bus every 5 min 6 am-6 pm M-Sat. Pay fare PP7 (30¢) before boarding. 20-60 min to Santa Cruz, Cubao, depending on traffic. Tends to be crowded, not much room for bags. OK if you know the territory.

Avis, Hertz, National.

P PP49 ($2)/day.

MARSEILLES, France
Aeroport Marseilles Provence, 17 mi (28 km) NW

F185 ($33.30) daytime, F256 ($46) night. 25 min to city. Baggage F3.50 (65¢) each. No tip necessary.

Aeroport Marseille Provence bus to and from Gare St. Charles every 20 min 6:20 am-10:20 pm. A 30-min trip. Fare F36 ($6.50). Service also to Aix-en-Provence, F29 ($5.20), 35 min.

Avis, Budget, Citer, Eurodollar, Europcar, Eurorent, Hertz.

P ST F47 ($8.45), LT F26 ($4.70).

MARTHA'S VINEYARD, Massachusetts
Dukes County Airport, 5 mi W of Edgartown, 5 mi S of Vineyard Haven

Hathaway Taxi: $10 flat rate for 2, $1 each addl, 15 min.

Hertz, National.

MATAMOROS, Mexico
Servando Canales Airport, 10 mi (16 km) S

$6, 25 min.

Avis, Budget, Dollar.

MAZATLAN, Mexico
Rafael Buelna Intl Airport, 14 mi (23 km) SW

$8-10, 25 min.

Budget, Avis.

MELBOURNE, Australia
Melbourne Intl Airport, 15 mi (24 km) NW

A$19 ($14.65), 20-35 min to Regent Hotel, Bourke Street area. No tip.

Skybus every 30-60 min 6:30 am-11:30 pm, A$6 ($4.60). Terminus is Skybus Travel Center, 58 Franklin St.

Avis, Budget, Hertz, Thrifty.

P ST A$10 ($7.70), LT A$4.50 ($3.50).

MEMPHIS, Tennessee

Memphis Intl Airport, 9 mi SE

$13-16, 20 min to Peabody, downtown. St. Jude Hospital $13, Germantown $16, NAS Memphis/Millington $32.50, Olive Branch $23, Southaven $13, W. Memphis $25.

Airport Limousine Service, $8. Use limo phone if none is waiting.

No. 32 bus at sign, lower level. Every 20-50 min 6:47 am-6:16 pm M-Sat. No Sun. 85¢ plus 10¢ transfer. Change at Fairgrounds to No. 56 or No. 10. Total time to 2nd & Madison downtown, 75-80 min.

Alamo, American International, Avis, Budget, Dollar, Florida, Hertz, National, Thrifty.

P ST $12, LT $6.

MERIDA, Mexico

Manuel Crescencio Rejon Airport, 5 mi (3 km) SW

$6, 15 min.

Avis, Hertz, Volkswagen.

MEXICO CITY, Mexico

Benito Juarez Airport, 6.2 mi (10 km) E of Zona Rosa

Buy trip ticket at Authorized Taxi Service booth. Fare to Zona Rosa $5-10 in airport cab, white and mustard colored. Four ride for price of one. Ride time 30 min. Tip not necessary but won't be refused. Sign in arrivals area cautions "Use only authorized service. The use of any other service is on your own responsibility," meaning it may cost double.

Luggage not allowed on Metro though a large briefcase, small shoulder bag probably OK. Metro is clean, fast, modern, cheap. Good way around in nonrush hours. Entrance marked by M on pylon to left outside terminal, a 3-4 min walk around parking garage. Fare P150 (5¢). Airport Station is on Line 5. To Zona Rosa take Pantitlan train. At Pantitlan (one stop), change to Observatorio train on Line 1. Get off at Insurgentes, in Zona Rosa.

Avis, Budget, Dollar, Hertz, National.

P ST $6.75, LT $4.95.

MIAMI, Florida

Miami Intl Airport, 6 mi (10 km) NW

Airport info by phone in Spanish & English: 305-876-7000.

$14, 20 min. Miami Beach $18-28, Key Biscayne $22, cruise ships $12, Doral Country Club, $13, Ft. Lauderdale-Hollywood Intl Airport $35. Flat fares to hotels in airport region range $5-8.

Northbound Tri-Rail rush hour AM & PM service M-F to West Palm Beach. Take shuttle from Concourse E, Lower Level. 15 min to RR station. Trains leave at 5:05, 6, 7 am; 3:45, 4:25, 5:15, 6, 7:20 pm. 1 hour 40 min to end of the line at W. Palm Beach with stops en route at Metrorail, Golden Glades, Hollywood, Ft. Lauderdale/Hollywood Intl Airport, Ft. Lauderdale, Cypress Creek, Pompano Beach, Deerfield Beach, Boca Raton, Delray Beach, Boynton Beach, Lake Worth, Palm Beach Intl Airport. Fare between any two stations $2. Parking & shuttles free. Info: 305-728-8445.

Southbound Airporter South to Homestead, Key Largo, Islamorada at 9:15 am; 12:30, 2:30, 5:30, 7:30 pm. Info: 305-247-8874.

SuperShuttle van service on demand. Downtown hotels, cruise ships, $7. Rates to other places: 305-871-2000.

No. 7 Metro bus to NW 2nd Ave. & 5th St. downtown. M-F 5:27 am-9:07 pm, 35-40 min downtown. $1 exact fare. To ride Metrorail, take transfer for 25¢. Hourly service Sat-Sun 6:25 am-5:16 pm. **To Miami Beach** No. J Metro bus every 20-30 min 4:39 am-11:31 pm; hourly Sun 5:40 am-8:40 pm. On Miami Beach, follows Collins Ave. route from 39th to 72nd St. $1.

Alamo, American Intl, Avis, Budget, Dollar, General, Hertz, Inter-American, Lindo's, National, Thrifty, U.S.A., Value, and others.

P ST $18, LT $6.

To Key West Greyhound at 7:10 am, 6:05 pm. Intermediate stops at Coral Gables, Homestead, Key Largo, Islamorada, Layton, Marathon, Tavernier, Perrine, Big Pine Key. To Key West $33 OW, $62.70 RT. Boards at bus loop, Lower Level. Info: 305-374-7222.

MIDLAND, Texas

Midland Regional Airport, 10 mi W

To Midland $12, 15-20 min. Fare slightly less to Odessa. Addl riders 25¢ each.

Avis, Budget, Dollar, Hertz, National.

MILAN, Italy

Linate Airport, 4.5 mi (7 km) E

Lit17,000 ($13.60), 15-20 min. Take yellow (or white with blue stripes) cab with meter, not a gypsy. Tip included in fare but bags are Lit500 (40¢) addl each.

STAM coach to Milano Centrale, largest railway terminal in the world. 7 am-11 pm at 15 min intervals. Fare Lit2500 ($2) purchased in advance at train station or from driver at airport. Luggage space beneath bus. 23-min ride. Bus also goes to Garibaldi Station.

ATM No. 73—look for sign and bus shelter near STAM. Every 12-23 min 5:35 am-12:50 am. Fare Lit1000 (80¢). Buy ticket from machine at bus stop. Bus terminates at Piazza San Babila near Duomo with connection to Milano Metro. No baggage space, likely to be crowded.

Avis, Budget, Europcar, Hertz, InterRent/Autotravel, Maggiore, Tirreno.

P Lit23,000 ($18.40)

MILAN, Italy
Milano/Malpensa Intl Airport, 28 mi (45 km) NW

Lit100,000 ($80), 40-60 min.

Airpullman meets intl flights 6:15 am-10:15 pm. Lit10,000 ($8). 60-min ride. Stops at Centrale, Garibaldi stations in Milan.

Bus to Gallarate RR Station then rail to Milan, Porta Garibaldi Station. Service approx hourly 6:10 am-6:50 pm. Total journey time 65-100 min.

Avis, Budget, Europcar, Hertz, InterRent/Autotravel, Maggiore.

P Lit18,000 ($14.40).

MILWAUKEE, Wisconsin
General Mitchell Intl Airport, 7 mi S

$14-16, 15-25 min downtown. Bayside, $28-31; Menomonee Falls, $31-37; Brookfield, $22-28; New Berlin, $17-22; Muskego, $20-25. Addl passengers 25¢ ea.

Downtown, $6.50. To west, north Milwaukee & suburbs, $9-15. To Sheboygan, Kohler: Limousine Services van, $20 OW. Info & res: 414-769-9100.

No. 80 bus every 15-50 min M-F 5:50 am-9:58 pm. Sat every 30 min from 7:09 am to 6:13 pm. Sun-Hol 11:50 am, 12:41 pm and approx every 30 min to 5:45 pm. $1 fare—bills OK. 35-min ride to 6th & Wisconsin downtown. Picks up at shelter at far right (north) end of ramp outside terminal.

To O'Hare Intl Airport, Chicago Greyhound. $13 OW, $20 RT. Info: 414-272-2954.

Alamo, Avis, Budget, Dollar, Hertz, National.

P ST $8, LT $4/3.

MINNEAPOLIS/ST. PAUL, Minnesota

Minneapolis-St. Paul Intl Airport, 13 mi (21 km) SE of Minneapolis, 8 mi (13 km) SW of St. Paul

To Mpls $20, 20-30 min. **To St. Paul** $13, 20-25 min. Others: Wayzata $30; Bloomington $14; 3M, $19; General Mills $26; Control Data $8.

To Mpls Airport Express every 15-20 min from lower level 5:30 am-midnight. $7.50 OW, $11.50 RT. 30 min to downtown hotels. **To St. Paul** Limousine Service every 20 min. $5.50-$8.50 to downtown, midway, East Side hotels. **To Eau Claire, Menomonie, Hudson** Eau Claire Passenger Service: 715-835-0399. **To Rochester** Jefferson Lines: 612-726-5118. Rochester Express: 507-288-4490. **To Marshall, Watertown, Glenwood, Morris, Rice Lake, Hayward, Ashland, Ironwood** Four Star Lines: 612-537-0604. **To Mankato** Mankato Land to Air: 612-625-3977.

To Mpls No. 7 bus from heated shelter alongside Green Concourse on the lower level. 5:09 am-12:10 am M-F every 20-40 min, every 60-min Sat-Sun-Hol. 90¢ peak times, 75¢ otherwise. 42 min to Hennepin & Washington downtown. No. 35P Express M-F to city at 6:38, 7:12, 7:26, 7:52 am; from city at 3:31, 4:35, 5:15, 5:32 pm. $1. To 2nd Ave. & 2nd St. **To St. Paul** Take No. 7 (above), transfer at Ft. Snelling to No. 9 for downtown. 30 min.

Alamo, Avis, Budget, Hertz, National have counters in terminal. American Intl, Dollar, IRA, Payless, Thrifty on call by courtesy phone.

P ST $24.25, LT $10. Econolot at 7100 34th Ave S., $6, free shuttle.

MOBILE, Alabama

Mobile Municipal Airport, 15 mi W

$15 for up to two people, 20¢ each addl. 20-30 min.

Mobile Bay Limousine 5 am-last flight. $10 to downtown hotels. AMEX.

Avis, Budget, Dollar, Hertz, National.

MOLOKAI, Hawaii

Molokai Airport, 8 mi NW of Kaunakakai

Gray Line: call 567-6177. To Ke Nani Kai, Paniolo Hale, Sheraton Molokai on west end, $6 per person. To Hotel Molokai, Pau Hana Inn on east end, $6 per person, minimum two.

Avis, Budget, Dollar, Tropical.

MOMBASA, Kenya

Moi Intl Airport, 8 mi (13 km) W

KS200 ($10), 15 min; to hotels, KS240 ($12), 20 min. Tip not necessary but KS20 ($1) OK for good service.

Kenya Bus Service departs terminal every 30 min 5 am-8 pm. Fare to driver KS6 (25¢). Baggage OK. 15 min to central city.

Avis, Diani, Eurocar, Hertz, Kenatco, Taxis Coop, Tsavo.

MONTEGO BAY, Jamaica
Sangster Intl Airport, 3 mi (5 km) N

To city center J$35 ($6.50), 10 min. To resort hotels J$67 ($12.50).

American Intl, Avis, Budget, Dollar, Greenlight, Hertz, Island, Jamaica, Liberty, National, Pleasure Tours, Thrifty, United. J$1 (20¢)

MONTEREY, California
Monterey Peninsula Airport District, 3 mi E

$6, 10-15 min. Hyatt, Holiday Inn, $4, 5 min; Doubletree, $6, 10 min.

Airport Limousine meets flights. Fares $4 and up depending on destination. Info: 408-372-5555.

No. 21 Salinas bus hourly M-F 7:28 am-6:51 pm. Sat-Sun 7:45 am-6:51 pm. $1.25. 15-min ride downtown.

American, Avis, Budget, Dollar, Hertz, National.

P ST $7, LT $5/day.

MONTERREY, Mexico
Mariano Escobedo Intl Airport, 12 mi NE

$7 flat rate, 30 min. Tipping not customary.

Avis, Budget, Dollar, Hertz, National, Thrifty.

MONTREAL, Quebec
Dorval Intl Airport, 12 mi (20 km) NW

C$26.30 ($22.35), 20-30 min.

Aerocar bus every 20-30 min 7 am-11:59 pm. C$7 ($5.80), 25 min to downtown hotels: Bonaventure, Le Chateau Champlain, Queen Elizabeth, Centre Sheraton.

Bus connection via No. 204 Cardinal Est (East), C$1 (85¢). Get a transfer. Change at Dorval station (a 5-min ride) to No. 211. 15-min express ride to Lionel Groulx station. Transfer to Metro for downtown. 45 min total.

To Mirabel Airport Aerocar hourly 10:20 am-10:20 pm. C$10 ($8.50), 50 min.

Avis, Budget, Empress, Hertz, Thrifty, Tilden.

P ST C$8 ($6.80), LT C$6 ($5.10).

Ski season special From Nov. 25 to March, Aeroski coach connects Dorval & Mirabel with Laurentians–Ste. Adele, Val David, Mt. Blanc, St-Jovite, Gray Rocks, Villa Bellevue, Auberge Cuttles, Manoir Pinoteau, Mont Tremblant. Info: 514-397-9999.

MOSCOW, Russia

Sheremetyevo Airport, 16 mi (26 km) NW

Rb10 ($16.50 at official exchange rate). 40 min to Kremlin–but only after waiting in cab line for up to 30 min. Tip 10%.

Hourly shuttle bus to Aeroport station on Metro Line 2. 5-kopeck coin (10¢) through turnstile. Metro runs 6 am-12:30 am. Five stops to Sverdlova Ploshchad–Red Square. Moscow Metro is fast, comfortable, frequent, cheap, and a celebration of Soviet architecture in marble and mosaic. No fun if you are laden with luggage, however.

Intourist makes arrangements.

MT. HAGEN, Papua New Guinea

Kagamuga Airport, 7 mi (11 km) NE

No taxis.

PMV–public motor vehicle–is a cross between minibus and truck. Not elegant but practical and cheap. 15-20 min, 30toea (35¢) to city center.

Avis, Budget.

MUNICH, Germany

Munich-Riem Airport, 6 mi (10 km) SE

DM22 ($13), 30 min to city center. Tip for good service a few marks.

Munich Airport City Service bus every 15 min 4:15 am-9 pm but meets flights outside these hours. DM5.50 ($3.30) to Hauptbahnhof (main station), where cabs are available. 20-25 min.

No. 91 from stop between terminals to Riem Station of U-Bahn (metro). Transfer to train S-1. DM3.60 ($2.10) bus fare includes transfer to U-Bahn.

Autohansa, Avis, Bach-Taxi, Europcar, Eurorent, Hertz, InterRent, Mages-Autovermietung, Schuldt, Sixt.

P ST DM25 ($14.75), LT DM11 ($6.50).

NADI, Fiji
Nadi Intl Airport, 5 mi (8 km) NE of Nadi, 90 mi (146 km) NW of Suva

F$5 ($4), 12 min. To Sigatoka, F$35 ($28), 2 hrs.

Courtesy service to major hotels.

Avis, A-Team, Budget, Hertz, Khans, National, Roxy, Thrifty.

NAIROBI, Kenya
Jomo Kenyatta Airport, 10 mi (16 km) SE

Ksh320 ($13), 15 min. Tip optional.

Kenya Bus Service No. 34 hourly 6:30 am-8 pm. Ksh8 (40¢). Baggage OK. 30 min to city center.

Across Africa Safaris, Ark Travel, Avis, Europcar (Airporter), Glory, Hertz/UTC, Kairi Tours & Safaris, Kenatco, Jambo Taxis, Payless.

NANTUCKET, Massachusetts
Nantucket Airport, 3 mi SE

$6, 7 min. To beach, $6, 5 min. Extra passengers, $1 each.

Budget, Hertz, National, Thrift.

P ST $10, LT $4/day.

NAPLES, Florida
Naples Municipal Airport, 3 mi NE

$6-8, 10 min to town. To Beach Club Hotel, $8.50, 10 min; Ritz-Carlton, $16, 20 min; Registry, $12, 15 min. Addl passengers $1 each. Baggage 50¢ each.

Alamo, Avis, Budget, Dollar, Hertz, National, Sears, Thrifty, Ugly Duckling.

P Free.

NAPLES, Italy
Capodichino Airport, 4 mi (7 km) N

Lit30,000-35,000 ($24-28) includes empty return, 30 min. Tip 10%.

Autotravel, Avis, Budget, Europcar, Euro-Rent, Eurotrans, Hertz, InterRent, Italy by Car, Maggiore.

P Lit12,000 ($9.60)/day.

NASHVILLE, Tennessee
Nashville International Airport, 10 mi SE

$12-14, addl riders 50¢ each. 15-30 min depending on traffic.

$8 OW, $14 RT to downtown hotels every 20-30 min 6 am-11 pm on Downtown Airport Express. **Opryland** shuttle every 30 min 6 am-11 pm. $10. **Clarksville/Ft. Campbell** Clarksville Limousine Service every 2 hrs 9 am-11 pm. $19 OW, $35 RT.

No. 18 Elm Hill Pike/Airport bus at 8:14, 9:34, 11:16 am; 1:37, 3:32, 5, 6:06, 7:35, 9:05, 10:35 pm. 36 min to Deaderick & 4th, downtown. 85¢. Boards outside lower level at MTA sign, extreme right.

Alamo, American Intl, Avis, Budget, Dollar, Hertz, National, Thrifty.

P ST $12, LT $5.

NASSAU, Bahamas
Nassau Intl Airport, 12 mi (19 km) SW

$10, 25-30 min to Nassau.

Avis, Budget, Hertz, National.

NEW ORLEANS, Louisiana
New Orleans Intl Airport, 14 mi W

$21 flat rate downtown for up to 2 people, $8 addl for 3 or more. 20-30 min.

Airport Shuttle, $7 downtown, French Quarter, 24 hrs, every 15 min. **To Slidell, Biloxi** Coastliner van at 8:30, 10:30 am; noon, 1:30, 3:30, 5:30, 7:30, 9:30, 11:30 pm. Slidell, $18 OW, $30 RT, 60 min; Biloxi, $29 OW, $50 RT, 2½ hr. Reservations required: (US) 800-647-3957, (Miss) 800-622-3022.

Airport-Downtown Express 5:30 am-11:30 pm M-F at 10-20 min intervals, 30-60 min until 11:30 pm. Sat-Sun-Hol on a 20-40 min schedule. $1.10 exact coins. 50-min ride to Elk Pl. downtown.

To Gulfport Municipal Airport See Coastliner listing at limos, above.

Alamo, American Intl, Avis, Budget, Dollar, Enterprise, General, Hertz, National, Payless, Thrifty, Value.

P $9.

NEW YORK, New York

Kennedy Intl Airport, 15 mi SE

Toll Free NY-Area number for Kennedy ground transportation information: 800-AIR-RIDE (247-7433).

The Port Authority advises: For ground transportation information and reservations go to the Ground Transportation Centers located in the baggage claim area of all terminals. Do not accept rides from anyone who approaches you in the terminal, unless you have already made prior arrangements.

$24-30 plus tolls, 35-60 min to Manhattan. To LaGuardia $15.

Carey Airport Express: $11-12.50 each way to Manhattan, $5 each way to Queens, $9.50 each way to Brooklyn. 718-632-0500/0509. Departs for Manhattan every 30 min from all airport terminals. Drop-off is at 125 Park Avenue, near the Grand Central Terminal between 41st and 42nd Streets (east side of Park Ave.) and the Port Authority Bus Terminal (42nd St. & 8th Ave.). Shuttle bus to the New York Hilton (West 53rd and 6th Ave.), Sheraton City Squire (7th Ave. between 51st and 52nd Streets), Marriott Marquis (Broadway between 45th and 46th Streets) and the Holiday Inn (Broadway and 48th St.) from Park Avenue available every 20 minutes 7:15 am to 10:15 pm daily. Travel time is approximately 60 minutes. Departs for Queens every 30 minutes from 5:30 am to 11 pm. Drop-off is at Long Island Rail Road Jamaica Station, Sutphin Blvd. and Archer Avenue. Travel time approximately 20 minutes. Departs for Brooklyn every 60 minutes from 8:30 am to 8:30 pm. Drop-off is at Ashland and Hanson Place near the Williamsburg Bank. Travel time is approximately 30 minutes.

Gray Line Air Shuttle: $14 per person 212-757-6840. Departs from all JFK terminals. Schedule varies according to passenger demand. Make arrangements at Ground Transportation Center or use courtesy phone to arrange pick-up. Drop-off anywhere between 23rd & 63rd Streets. Travel time is approximately 55 minutes.

New York Helicopter: $65, plus tax, one way* 800-645-3494. Frequent departures throughout the day from TWA International Terminal/Gate 37 to East 34th Street heliport. Travel time is approximately 15-20 minutes.
* Reduced rate available for connections with some airlines.

JFK Express: No longer in operation; discontinued April 1990.

Take free Long Term Parking Lot bus from each terminal every 10-15 min 24 hours to Howard Beach subway station. Connects with the A train Far Rockaway line to Manhattan. About 70 min ride to Midtown. Train fare $1.15 or token.

Q10/Green Bus Lines: 718-995-4700. Local service to Queens (Lefferts Blvd. and Kew Gardens) for connections with NYC subway system. (A,E,F and R trains.) Fare is $1.15 (exact change required).

Q3/N.Y.C.T.A.: 718-330-1234. Local service to Queens (169th St. and Hillside Ave.) for connections with NYC subway system (F and R trains) other local bus services along its route and Long Island Rail Road, Hollis or Locust Manor Stations. Fare is $1.15 (exact change required).

To LaGuardia Carey Transportation: 718-632-0500/0509. Scheduled bus service every 30 min to LaGuardia Airport. Fare is $9.50 **To Newark Intl** Princeton Airporter: 609-587-6600. Scheduled service to Newark Airport. Fare is $19.

Avis, Budget, Dollar, Hertz, National.

P ST $24, LT $5. Free, 24-hour shuttle every 10 min peak hours, every 30 min late night to remote lot.

For services to Long Island, Connecticut, New Jersey Make arrangements at JFK Ground Transportation Center.

NEW YORK, New York

LaGuardia Airport, 8 mi NE

Toll free NY-area number for LaGuardia ground transportation information: 800-AIR-RIDE (247-7433).

The Port Authority advises: For ground transportation information and reservations go to the Ground Transportation Centers located in the baggage claim area of all terminals. Do not accept rides from anyone who approaches you in the terminal, unless you have already made prior arrangements.

$14-20 to mid-Manhattan plus tolls. 20-40 min. To JFK, $16-20 plus tolls, 20-30 min. To Newark Intl Airport fare is metered amount—$40-45—plus $10 surcharge, plus tolls. To Westchester, Nassau counties, metered fare to NYC limit, then double meter for remainder.

Carey Airport Express: $8.50-10 each way to Manhattan, $5 each way to Queens, $7.50 each way to Brooklyn. 718-632-0500/0509. **Departs for Manhattan** every 20-30 min from 6:45 am to midnight. Drop-off is at 125 Park Avenue, near Grand Central Terminal between 41st and 42nd Streets (east side of Park Ave.), the New York Hilton (West 53rd Street and 6th Ave.),

Sheraton City Squire (7th Ave. between 51st and 52nd Streets), Marriott Marquis (Broadway between 45th and 46th Streets), and the Holiday Inn (Broadway and 48th Street) from Park Avenue available every 20 minutes 7:15 am to 10:15 pm daily. Travel time is approximately 60 minutes. **Departs for Queens** every 30 minutes from 5:30 am to 11 pm. Drop-off is at the Long Island Rail Road Jamaica Station, Sutphin Blvd. and Archer Avenue. Travel time is approximately 20 minutes. **Departs for Brooklyn** every 60 minutes from 9:15 am to 9:15 pm. Drop-off is at Ashland Place and Hanson Place near the Williamsburg Bank. Travel time is approximately 30 minutes

Gray Line Air Shuttle: 212-757-6840. $11 per person. Departs from all LaGuardia terminals. Schedule varies according to passenger demand. Make arrangements at a Ground Transportation Center. Drop-off anywhere between 23rd and 63rd Streets. Travel time is approximately 45 minutes.

Q-33 Bus: 718-335-1000. Fare is $1.15 (exact change in coins required). Operated by Triboro Coach for service to Queens and connections to subway (E,F,G,R and #7 trains). Departs every 10-20 minutes, 24 hours a day.

Q-48 Bus: 718-330-1234. Fare is $1.15 (exact change in coins required). Operated by Metropolitan Transportation Authority for service to Queens and connections to the Long Island Rail Road, Flushing Station. Departs every 15 minutes from approximately 5:30 am to 12:30 am.

To Wall St. Pan Am Water Shuttle, M-F only. Take free Pan Am shuttle bus every 15 min to boat dock at Marine Air Terminal. 40 min to Pier 11, corner Wall & South Sts. Intermediate stop at 35th St. & East River. $25 OW, $45 RT. Info: 800-543-3779.

To Kennedy Intl Airport Carey Transportation: 718-632-0500/0509. Scheduled bus service every 30 minutes to JFK. Fare is $9.50. **To Newark Intl Airport Via Manhattan-Transfer Required** Carey Transportation: 718-632-0500/0509. Departs for Manhattan every 20-30 minutes. Drop-off is at 125 Park Ave. between 41st and 42nd Streets or Port Authority Bus Terminal's Airport Bus Center, 41st Street and 8th Ave. Fare is $8.50. TRANSFER TO Olympia Trails: 212-964-6233. Departs for Newark Airport every 20-30 min 5 am-11 pm. Pick-up is at Park Ave. and 41st Street. Fare is $7. OR N.J. Transit: 201-460-8444. Departs for Newark Airport every 15-30 minutes, 24 hours. Pick-up is at Port Authority Bus Terminal's Airport Bus Center, 41st Street and 8th Ave. Fare is $7.

Avis, Budget, Dollar, Hertz, National.

P $4 first 4 hours or part, $15/day.

To Long Island Classic Share Ride/Limousine shared door-to-door service. Fares range from $26 to Valley Stream to $50 to Riverhead. Info: 516-567-5100. **To Connecticut** Connecticut Limousine. Fares $18-33. Info: 800-242-

2283 (NY) or inquire at Ground Transportation Center. **To Westchester County and Upstate New York** Connecticut Limousine, 800-243-6152, 914-699-1000; Westchester Express door-to-door, 914-667-2400, fares $20-40. For services to New Jersey make arrangements at LGA Ground Transportation Center.

NEWARK, New Jersey

Newark Intl Airport, 16 mi SW of midtown New York City

Toll Free NY-NJ Area Number for Newark ground transportation information: 800-AIR-RIDE (247-7433).

The Port Authority advises: For ground transportation and reservations go to the Ground Transportation Centers located on the lower (baggage claim) level of all terminals near exit Door 2. It is not advisable to accept rides from anyone who approaches you in the terminal, unless you have already made prior arrangements.

$30 plus tolls, 30-45 min to Manhattan from Battery to West 59th St. Flat rates: To LaGuardia $44.50 plus tolls, JFK $53.50 plus tolls, Atlantic City $131, Basking Ridge $41, Clifton $28, Elizabeth $14, Red Bank $54, Ridgefield $31, Summit $24, Parsippany $38, Oradell $35, Hoboken $23.

Olympia Trails Express Bus: 201-589-1188 or 212-964-6233. Fare is $7. Departs for downtown Manhattan (World Trade Center) 7 am to 10 pm and midtown Manhattan (Grand Central Terminal at 41st St. & Park Ave., or Penn Station at 34th & 8th Ave.) about every 20 minutes, 6:15 am to midnight. Travel time to downtown 20 to 40 minutes, to midtown 30 to 50 minutes.

NJ Transit No. 300 Express to Port Authority Bus Terminal, 8th Ave. & 42nd St. Every 15-30 min 24 hrs a day. 30-45 min travel time. Fare is $7.

Gray Line Air Shuttle to any Manhattan location between 23rd & 63rd St. Departures vary with demand. Fare $16. Make arrangements at Ground Transportation Center.

Airlink/NJ Transit #302: 201-460-8444 or 800-772-2222 (NJ only). Fare is $4. Departs for Newark's Penn Station for rail connections on NJ Transit, AMTRAK and PATH. Operates 20 hours daily starting at approximately 6 am.

PATH trains from Track 1 Newark Penn Station to lower Manhattan. $1 fare, coins or bill, in turnstile. Very frequent service. Info: 800-234-7284. NJ Transit or Amtrak trains to **Penn Station, Manhattan** from **Newark Penn Station.** NJ Transit $2.50. AMTRAK $7.

To JFK airport: Scheduled service is provided by Princeton Airporter. Info: 609-587-6600. Fare is $19.

Avis, Budget, Dollar, Hertz, National.

P ST $4/hr first 4 hrs or part; penalty $48 daily rate. LT $22, $6. Parking for up to 24 hrs can be prepaid in baggage claim, terminals A, B, C.

To New Jersey Airport Limousine Express: 201-621-7300 or 800-624-4410 (NJ only). Scheduled service to Fort Monmouth, Middlesex, Monmouth and Union counties. Fares range from $13 to $20.
NJ Transit: 201-460-8444 or 800-772-2222 (NJ only). Scheduled bus service to Fort Dix, McGuire Air Force Base, Downtown Newark and Elizabeth, Essex, Union, Somerset and Hunterdon counties. Fares range from $1.15 to $5.30.
Princeton Airporter: 609-587-6600 or 800-451-0246 (NJ only). Scheduled service to Middlesex, Morris and Mercer counties. Fare is $14 to $18. Trans-Bridge Bus Lines: 215-868-6001. Scheduled coach service to Hunterdon County. Fare is $9.20 to $11.30.
To Pennsylvania Princeton Airporter: 609-587-6600 or 800-451-0246. Scheduled service to Yardley, Pennsylvania. Fare is $18.
Trans-Bridge Bus Lines: 215-868-6001. Scheduled coach service to Easton, Bethlehem and Allentown, Pennsylvania. Fares range from $10.50 to $11.30.
For services to Pennsylvania, Long Island, Connecticut, Upstate New York make arrangements at EWR Ground Transportation Centers.

NICE, France

Nice/Cote d'Azur Airport, 4 mi (7 km) W

F80-100 ($14.40-18) to city center for 3 passengers, 4 only if driver agrees. Extra for luggage F5 (90¢) apiece. Tip 10-15% for special service. 10-30 min depending on traffic. Surcharges Sun, Hol. To Cannes, FF350 ($63).

Special Aeroport express runs along Promenade des Anglais coast road. Look for BUS signs on wall at west Terminal 1 exit. Pay fare F19 ($3.40) aboard. Service from 6:10 am (7:10 Sun) to 11:10 pm, at 20-25 min intervals peak hours, 30-60 min other times. Small bus, little room for luggage. Some services call at SNCF railway station in Nice by request. Gare Routiere Departmentale (Central Coach Station) is end of the line. 15-35 min.

No. 23 bus stops across road and to left of express coach described above. Pay F8 ($1.45) aboard. No baggage space and may be quite crowded during peak hour commutes. Follows inland route, numerous stops. 45-50 min to Place Massena & Ave. Jean Medecin. Inbound, bus is marked PORT; outbound, AEROPORT.

Avis, Budget/Milleville, Citer, Europcar, Eurorent, Hertz, InterRent, Mattei, Sporting Intl.

P F54 ($9.70).

NORFOLK, Virginia

Norfolk Intl Airport, 6 mi NE

$15, 15-25 min to Norfolk; $28, 20 min to Virginia Beach. To Williamsburg, $70 flat for up to 5 people, 45 min.

Airport Limousine Service, $7.50 to Norfolk; $12 to Virginia Beach; $19 to Williamsburg. Hourly service every day to last flight.

Avis, Budget, Dollar, Hertz, National.

NUREMBERG, Germany

Nuremberg Airport, 4.3 mi (7 km) N

DM25 ($15), 15-20 min to city center. Same fare & travel time to Fair Ground, Meistersingerhalle/Convention Center. Tip 10%, rounding up to next mark.

Neukam-Romming shuttle every 45 min 7:55 am-10:50 pm M-F. Sat: 9, 9:45, 10:25 am; 2:15, 5:45, 7:30, 10:50 pm. Sun: 10:30 am; 12:30, 2:15, 3, 6:15, 7:30, 9:10, 10:50 pm. DM5 ($3). 15-min to Hauptbahnhof.

Service from front of airport every 30 min 6 am-11 pm. 40-60 min to city center. Fare DM2.80 ($1.70). Baggage OK.

Auto Hansa, Avis, Europcar-National, Hertz, InterRent, Sixt-Budget.

P DM16 ($9.60)/day.

OAKLAND, California

Oakland Intl Airport, 11 mi SE of Oakland, 19 mi SE of San Francisco

Oakland, $16-17, 10-15 min. San Francisco, $30-35, 25-60 min plus $1 bridge toll Sun-Th, $2 F-Sat.

Door-to-door van service available from AM/PM Airporter (415) 547-2155; Bayporter Express (415) 467-1800; Direct Shuttle (415) 674-0474; Express Shuttle (408) 378-6270; North Bay (800) 675-1115 (from CA); South Bay Airport Shuttle (408) 559-9477.

Take Air-BART shuttle from shelter on center island to left outside entrance. Every 10 min, 6 am-midnight M-Sat (from 9 am Sun). Baggage rack inside. $2, 10-min ride to Coliseum station of BART. To Oakland City Center-12th St. take Richmond train. 80¢, 11 min. To downtown San Francisco, take Daly City train, $1.90, 21-27 min to Embarcadero, Montgomery, Powell, Civic Center stations. BART is clean, quiet, fast. Plenty of information, easy to use.

AC Transit No. 58 to downtown Oakland M-F every 30-60 min 5:51 am-7:42 pm. $1, 30 min to 12th & Franklin.

To San Francisco Intl Airport (SFO) Bay Porter Express hourly on the hour 6 am-11 pm. $10, 60 min to SFO. Each run begins at Parc Plaza, Oakland, 15 min before Oakland departure.

A-1, Alamo, Avis, Budget, Dollar, General, Hertz, National, Thrifty, U-Save.

P ST $15, LT $6/$4. Free shuttle every 5 min.

To Travis AFB, Fairfield Travis/Solano Airporter. 707-437-4611. **To Pleasanton, Dublin, San Ramon, Castro Valley** San Ramon Valley Airporter Express. Res: 415-484-4044. **To Napa, Vallejo** Grapevine Airport Service. Res: 707-253-9093.

OKLAHOMA CITY, Oklahoma

Will Rogers World Airport, 11 mi SW

$12, $1 each addl rider. 20 min. **To Norman** $22-30, 30 min.

Airport Express shared-ride vans. $6 first person, $3 each addl to downtown OK City hotels. **To Norman** $22.50, $3 each addl. 40 min. Leaves from lower level adjacent to baggage claim. AMEX, MC, VISA.

Avis, Budget, Dollar, Hertz, National, Thrifty.

P ST $8, LT $4-5/day, $20-25/week.

OMAHA, Nebraska

Eppley Airport, 5 mi NE

$5-6, 5-10 min.

$5-6. Call 342-1131 if station wagon is not at airport.

Alamo, American Intl, Avis, Budget, Dollar, Hertz, National, Thrifty.

P ST $12, LT $3.

ONTARIO, California

Ontario Intl Airport, 25 mi SW of San Bernardino; 19 mi NW of Riverside; 10 mi E of Pomona; 39 mi E of downtown Los Angeles

Pomona $20, 25 min; Riverside $32-35, 40 min; San Bernardino/ Loma Linda $40, 45 min; Los Angeles (LAX) $85-90, 75-90 min. **To downtown LA** SuperShuttle $50.
2No. 496 RTD bus loads to right along sidewalk past car return. Eastbound service to San Bernardino/Riverside. Westbound to LA. M-F at 5:58, 7:03, 8:22 am, then hourly at :36 to 3:36 pm; then 4:38, 5:38, 6:38, 7:33 pm. 75-min run to 7th & Maple, downtown LA. $3.70. Slower service to LA on RTD No. 484, $2.90. To Disneyland, Knotts Berry Farm, RTD No. 149.

To LAX SuperShuttle $21. Info & res 800-554-6458.

Avis, Budget, Dollar, Hertz, National.

P ST $8, LT $3.

OPORTO, Portugal See PORTO, Portugal

ORANGE COUNTY, California

John Wayne Airport, 5 mi S of Santa Ana; 5 mi NE of Newport Beach; 10 mi SE of Disneyland/Anaheim Convention Center

Airport info 714-755-6500

Newport Beach, $14-16; Anaheim/Disneyland, $24-25; U.C. Irvine, $7; Santa Ana, $14; Costa Mesa, $6-$8; Los Angeles Intl (LAX), $65.

To Disneyland/Anaheim Amtrans, SuperShuttle, and other van services: $10 per person to hotels. Airport Coach every 45 min. $7 adults, $5 kids. **Orange County door-to-door** Fares vary. Know ZIP code of your destination for fastest info. Amtrans 714-220-1122. SuperShuttle 714-973-1100. **To LAX** Amtrans, SuperShuttle vans, $15 per person. AMEX, MC, VISA accepted.

Avis, Budget, Dollar, General, Hertz, National.

P ST $14, LT $7/day.

ORANJESTAD, Aruba

Queen Beatrix Intl Airport, 2 mi (1.25 km) E

$8 flat rate, 7-10 min. To hotels on W coast, $8, 15-20 min. Baggage $1. Tip $1-2. (US currency is as readily accepted as the Aruban florin.)

ARUBUS picks up 5-min walk from airport, hourly 5 am-11 pm. 90¢ 10-15 min to Oranjestad. Small baggage OK.

Avis, Budget, Dollar, Hertz, National, Thrifty.

P $4.60/day.

ORLANDO, Florida

Orlando Intl Airport, 12 mi SE of Orlando, 23 mi NE of Walt Disney World.

$23.50, 20 min to Orlando; $36.50, 35 min to Walt Disney World.

Mears Shuttle, Transtar Shuttle, $10 OW, $17 RT downtown. To Walt Disney World, $12.50 OW, $21 RT adult; $9.50 OW, $15 RT child. AMEX, VISA, MC accepted at airport counter.

No. 11 bus hourly 6:30 am-11:20 pm. 75¢. 40 min to Pine St. terminal, downtown.

Avis, Budget, Dollar, Hertz, National, Payless.

P $6/day.

OSAKA, Japan
Osaka Intl Airport, 10 mi (17 km) NE

Y4500 ($29.25), 30-45 min downtown. To Kyoto, Y13,700 ($89), 50 min; to Kobe, Y8250 ($54).

Osaka Airport Transport Co. every 15 min 8 am-8 pm. Y380 ($2.50). 35-45 min downtown Osaka. To Kyoto, Y770 ($5.20); to Kobe, Y620 ($4).

Avis, Hertz.

OSLO, Norway
Fornebu Airport, 6.25 mi (10 km) W

NKr100 ($15), 15 min. Tip 10%.

SAS bus from outside arrivals hall to Central Railway Station (harbor side), Hotel Scandinavia. Departures every 15 min 6 am (8 Sun) to 11:30 pm. Pay driver NKr30 ($4.50). 20-25 min ride. On Sat evening, Sun, bus runs every 30 min. Info phone: 59 68 14.

Bus No. 31 outside arrival hall. NKr17 ($2.55) to driver. Departures every 15-30 min 6:08 am-12:08 am. No special baggage space. Many stops into town: 30-35 min ride.

Avis, Budget, Europcar, Hertz, InterRent.

P NKr70 ($10.50)/day.

OSLO, Norway
Gardermoen, 33 mi (53 km) NE

NKr450 ($67.50), 50 min.

Meets flights. NKr60 ($9).

Avis, Budget/Hasco, Europcar, Hertz/Kjoles, InterRent

OTTAWA, Ontario
Ottawa Intl Airport, 5 mi S

C$19 ($16.70), 20-30 min.

Pars Hotel Shuttle, C$8 ($7) every 25 min 6 am-11 pm to downtown hotels

Avis, Budget, Hertz, Tilden.

PALERMO, Sicily
Punta Raisi Airport, 18 mi (30 km) NW

Lit50,000 ($40), 40-50 min.

Hotels Porto Rais, Azzolini Residence, Palm Beach provide free transport

Prestia e Comande bus every 60-90 min 5:45 am-last flight. 50 min to Politeama Square. A/C. Baggage OK. Fare to driver Lit4200 ($3.35).

Avis, Europcar, Hertz, Holiday, InterRent, Italy by Car, Maggiore.

P Lit10,000 ($8) first day, Lit6,000 ($4.80) succeeding days.

PALM SPRINGS, California
Palm Springs Regional Airport, 2 mi E

$5, 5 min. To Rancho Mirage $27, 15 min. To Palm Desert $32, 20 min.

Sunbus, from road in front of airport property. Service 6 am-6 pm. every 30 min. Pay driver 50¢. 15 min into city. "Courtesy stops anywhere upon request."

Alamo, Avis, Budget, Dollar, General, Hertz, National, Payless, Thrifty, and others.

P $5.50/day.

PALMA, Mallorca
Palma de Mallorca Airport, 7 mi (11.3 km) SE

Pta1150-1265 ($9-10), 20 min.

No. 17 every 30-60 min 7:05 am-12:05 am. 30 min to Plaza de Espana, city center. Fare Pta120 (95¢) days, Pta135 ($1.05) nights.

Europcar, Hertz, InterRent/Atesa, Regent.

PANAMA CITY, Florida
Bay County Airport, 5 mi NW

$7, 15 min. To Panama City Beach, $17.

Avis, Budget, Dollar, Hertz, National, Snappy, Thrifty.

PAPEETE, Tahiti

Faa Airport, 2.5 mi (4 km) W

F1300 ($14), 7 min. Beachcomber Hotel, F500 ($5.40), 3 min; Maeva Beach Hotel, F750 ($8), 4 min; Tahara'a Hotel, F2000 ($21), 25 min. Baggage F50 (50¢). No tip.

Le Truck—a jitney with wooden benches—provides colorful, cheap ride into Papeete or anywhere else on the island. To town, about 100 francs (90¢), kids half fare.

Avis, Budget, Hertz, Pacificar, Roberts.

PARIS, France

Charles de Gaulle Airport (Roissy), 15 mi (25 km) NE

F160 ($27.85). After 8 pm, F180 ($31.30). F4 (70¢) each bag. 45-60 min ride. Tip not compulsory but 10% OK. Most cabs have a limit of 3 passengers. Add F5 (90¢) for a fourth.

Air France bus from Aerogare 2B, Gate B6; Aerogare 2A, Gate A6; Aerogare 2D, Gate D7. To Etoile every 15 min 5:40 am-11 pm, a 30-min ride. F38 ($6.85). Party of 3, F90 ($14.80), party of 4, F112 ($20). To Porte Maillot every 20 min 5:40 am-11 pm, a 30-min ride. F38 ($6.85). Party of 3, F90 ($16.20), party of 4, F112 ($17). To Gare Montparnasse (connection with TGV-very fast train) every hour, 7 am-7:30 pm. F62 ($11.15), party of 3 F140 ($25.20), party of 4 F165 ($29.70).

No. 350 to Gare du Nord, Gare de l'Est, No. 351 to Place de la Nation. Stops Aerogare 1 at RATP sign on Boutiquaire level; Aerogare 2A Gate A5; 2B Gate B6. Fare 6 Metro tickets, F31.20 ($5.60) purchased aboard. Packets of 10 Metro tickets purchased in advance also F34.50 ($6.20). No. 350 every 15 min, 6 am-11:50 pm; No. 351 every 30 min, 6 am-9:15 pm. Many stops, little bag space.

Shuttle bus every 5 min to Roissy Rail from Gate 28 Aerogare 1; gates A5, B6, D6, Aerogare 2. Buy rail ticket at Roissy Rail Station. Trains to Gare du Nord every 15 min 5 am-12 pm. 2nd Class F29 ($5.20); 1st Class F43.50 ($7.85). Comfortable, plenty of baggage space. A 35-min ride.

To Orly Air France bus from Gate 36, Aerogare 1; Gate A3, B10, Aerogare 2. Every 20 min, 6 am-11 pm, F63 ($11.35). 75-min ride.

Avis, Budget, Citer, Eurodollar, Europcar, Eurorent, Hertz, Thrifty.

P ST F60 ($10.50), LT F35 ($6.10). Weekend rate from 1:30 pm Fri to 1 pm Mon, F120 ($21). Free shuttle LT parking from Gate 30 Aerogare 1, gates A5, B6 Aerogare 2.

PARIS, France

Orly Airport (Sud and Ouest), 8 mi (14 km) S

F130 ($22.60) in heavy traffic. F4 (70¢) per bag. 40-45 min to Opera. Most cabs have a limit of 3 passengers. Add F5 (90¢) for a fourth. Tip not compulsory but 10% OK.

Air France coach, F31 ($5.60). Party of 3, F80 ($14.40); party of 4, F100 ($18). Board at Orly Sud Gate J, Orly Ouest Gate E, arrival level. Service every 12 min 5:50 am-11 pm. 45 min to Porte d'Orleans, Gare Montparnasse, where there are Metro connections; then Gare des Invalides, connection to RER rail. Comfortable, baggage space, available to passengers on all airlines. Alternative: Jet Bus Coach. Board at Orly Sud Gate E, Orly Ouest Gate C, arrival level. Every 12 min 5:48 am-11:36 pm. 12-min ride to Metro station Villejuif Louis Aragon (Line 7). F17 ($3.05).

ORLYBUS to Place Denfert-Rochereau, a Metro connection. Every 15 min 6:30 am-10 pm, then 10:30, 11, 11:30 pm. Fare F21 ($3.80), 40-min run. Travelers with light luggage who know Paris will find this service a way to avoid possible traffic delays on Air France bus to Invalides. Departs Orly from Porte F, Sud Terminal; Porte R, Ouest Terminal.

Take shuttle to Orly Rail from Sud Gate H, Ouest Gate F, arrivals level. Buy ticket at rail station. To Paris, Metro included: 2nd Class F23 ($4.15), 1st Class F33.70 ($6.05). Shuttles frequent, trains every 12-35 min 5:35 am-11:15 pm to Gare d'Austerlitz with several stops en route. Comfortable, ample baggage space.

ORLYVAL automated Metro Shuttle bus to Gare d'Antony from Sud Hall 2, Ouest Hall 3. Every 4 to 7 min, 6 am-11:50 pm. Fare: F55 ($9.90). Packet of 9 tickets: F49 ($8.80).

To Charles de Gaulle Airport Air France bus departs Sud Porte B, Ouest Porte D every 20 min 6 am-11 pm. F63 ($11). A 75 min connection.

Avis, Budget, Citer, Eurodollar, Eurorent, Europcar, Hertz, Thrifty.

P ST F60 ($10.50), LT F35 ($6.10). Free shuttle from Sud Porte H, Ouest Porte F.

PENANG (GEORGETOWN), Malaysia

Penang Intl Airport, 11 mi (17.6 km) SW

RGT15.40 ($5.55), 35 min.

No 83 (yellow bus) at terminal bldg., hourly on the hour, RGT1.15 (40¢).

Avis, Budget, Hertz, National, Sintat, and others.

PENSACOLA, Florida
Pensacola Regional Airport, 5 mi N

$7, 50¢ each addl passenger. 15 min.

Avis, Budget, Dollar, Hertz, National, Snappy, Thrifty.

PEORIA, Illinois
Greater Peoria Regional Airport, 5 mi SW

$9.15, flat rate, addl passengers 20¢ each. 20 min.

Avis, Budget, Dollar, Hertz, National.

PERTH, Australia
Perth Airport, 7 mi NE

A$18 ($13.90) from intl terminal; A$12 ($9.25) from domestic terminal. 15-20 min. To Fremantle, A$28 ($21.60), 45 min; Scarbrough, A$27 ($20.80), 45 min. South Perth, A$15 ($11.60), 20 min. Tip: Passenger discretion.

Airport Bus meets flights. A$6 ($4.60) to hotels in city center. 15-20 min from intl terminal.

Avis, Budget, Hertz.

P A$6 ($4.75)/day.

PHILADELPHIA, Pennsylvania
Philadelphia Intl Airport, 8 mi SW

$22-25, addl riders 20¢ each. 15-30 min.

Door-to-door shared ride. Look for "LIMO On Demand Service" outside baggage claim. $10 to Center City hotels. 15-30 min.

To Cherry Hill, Atlantic City Rapid Rover Airport Shuttle. To Cherry Hill $15.50; Atlantic City $52.50, 3+ passengers $17.50 each. Info: 609-428-1500. **To Wilmington area** Wilmington Shuttle, 302-655-8878. Downtown fare, $15.50.

Serving all Phila. areas and suburbs: Philadelphia Airport Shuttle, Inc. 215-969-1818. Outside PA 1-800-642-0080.

AIRPORT rail to Center City. Well marked—follow signs. Every 30 min 5:30 am-11:30 pm. $4.50. Stops at **30th St. Station:** Amtrak connections to Trenton, New York City, Harrisburg, Pittsburgh, Wilmington, Baltimore, Washington. Station is near U. of Penna., Drexel, Civic Center. **Penn Center:** Business district, hotels, City Hall. **Market East:** Historic district— Independence Hall, Liberty Bell, Betsy Ross House. **Temple U, Broad St. North** Travel time airport-Penn Center, 25 min.

Alamo, Avis, Budget, Dollar, Hertz, National.

P ST $30, LT $14, Remote Economy $6. Shuttle every 5 min.

PHOENIX, Arizona
Phoenix Sky Harbor Intl Airport, 4 mi SE

$7-9, 10-15 min to downtown Phoenix; $14-16 to downtown Scottsdale; $16-18 to downtown Mesa.

SuperShuttle shared-ride van service. Typical rates: Phoenix downtown $5, Sun City $12-20, Mesa $12-20, Scottsdale $10-20, Tempe $5. Confirm with driver. Boarding areas outside baggage claim Terminal 4 and at ground transportation building between Terminal 2 & Intl Terminal. Info & res: 800-331-3565. AMEX, VISA, MC accepted. Limousine service, 1-2 passengers: downtown, $15; Scottsdale, $20-30; Sun City, $35-40.

No. 2 bus every 30 min 5:59 am-8:37 pm from Terminal 3, Executive Terminal. 24 min downtown. 75¢ exact change.

Ajax, Alamo, American Intl, Avis, Budget, Dollar, General, Hertz, National, Payless, Rent-a-Dent, Value.

P ST $10, LT $3. Free shuttle to remote lot every 10-15 min, 30 min late night.

PITTSBURGH, Pennsylvania
Greater Pittsburgh Intl Airport, 17 mi W

$25, 25-40 min.

Airlines Transportation bus from lower level near United baggage claim. Departures every 20-30 min weekdays; up to 60 min late night, Saturdays. $8 to Westin Wm. Penn, Hilton, Sheraton, Hyatt, Vista Intl. Also serves Oakland, Mt. Lebanon, Monroeville. Info: 412-471-8900.

Avis, Budget, Hertz, National, Payless, Snappy, Thrifty.

P ST $30, LT $6.

POINTE-A-PITRE, Guadeloupe
Aeroport Pointe-a-Pitre/Le Raizet, 1.8 mi (3 km) N

F29 ($6), 10 min; to Gosier, F69 ($14), 20 min; to Sainte-Anne, F167 ($35), 35 min; to Bas du Fort, F50 ($10), 15 min; to St. Francois, F167 ($35),

5 min. 40% surcharge 9 pm-7 am. Baggage F1.5 (25¢) after first piece; ulky items 3F (50¢) each.

Co-Transport Society every 20 min 6 am-7 pm. 5F ($1). 20 min to ointe-a-Pitre. Baggage OK.

Avis, Budget, Carpentier, Europcar, Guadeloupe Cars, Hertz, InterRent, umbo, Karukera, Soltour.

F6 ($1.20)/day.

ORT MORESBY, Papua New Guinea

ackson Airport, 5 mi (8 km) E

Metered taxis: K12 ($11), 15-20 min downtown; K5 ($5.50) to Boroko; 7 ($7.70) Waigani. Alternative: PMV—public motor vehicle. Hybrid minibus/ uck with wooden benches. They wait on road across parking lot outside erminal. Any trip in Port Moresby costs 40toea (45¢).

Avis, Budget.

ORT-AU-PRINCE, Haiti

ort-Au-Prince Intl Airport, 4 mi (6.4 km) NE

$10, 20 min. To Petion-ville $12, 30 min; Carrefour $14, 45 min. Tip 10%.

Royal Haitian has a free airport van.

Avis, Budget, Hertz, J&S, National, Secom, Toyota.

ORTLAND, Maine

ortland Intl Jetport, 2 mi SW

$8, 10 min to intown Portland.

The Metro every 30 min 6:55 am-10:15 pm. 80¢ exact change. No Sun ervice.

Ajax, American International, Avis, Budget, Hertz, National, Thrifty.

ST $15, LT $7.

ORTLAND, Oregon

ortland Intl Airport, 9 mi NE

$19-22 downtown, 20 min. Extra passengers, 50¢ each. **Ground trans- ortation info booth** in center of covered waiting area.

Raz Tranz Airporter at :15, :35, :55 weekdays, :05, :35 weekends, 5:35-12:05 am. $6 to Convention Center, King's Way, Greyhound, Benson, Hilto Heathman, Portland, Marriott. 20-40 min. Service to other hotels available subject to space/ time. Check with driver. **To hotels/businesses within 4 m of airport:** Bilix Short Shuttle every 20 min 8 am-12:20 am. $4-7. **To Vancouver:** Vancouver Airporter hourly 7:05-12:05 am. Downtown, $7.50; Mar 205 Motor Inn, $5; Hazel Bell, $10. 573-9412. **To Gresham, Clackamas, Ea Multnomah County:** Eastside Airporter hourly 7 am-12 am. $10. 2-3 passengers to same address, $15. **To Beaverton, Washington Sq, King City, Tigard, Lake Oswego, Tualatin, Wilsonville:** Wilsonville/Tualatin Airporter every 30-60 min 9 am-11:30 pm. Info: 692-5222. Beaverton only: Beaverto Airporter at :15, 7:15-12:15 am. 649-2213.**To Salem** Hut Limo, 363-8059 **T Bend, Redmond** Central Oregon Shuttle. Info: 382-9371.

Tri-Met No. 12 Sandy Blvd. bus every 15-30 min 5:33 am-11:50 pm. Fir bus Sunday 6:21 am. Pay driver 90¢. Running time to Portland Mall, downtown, 39 min. Bus connects with "Max" trolley downtown & Gresham at Hollywood Transit Center.

Alamo, American Intl, Avis, Budget, Dollar, Hertz, National, Park-Shuttle-Fly, Thrifty.

P ST $15, LT $5.50/3.50. Shuttle every 5 min 5:30 am-11 pm, then every 10 min. MC, Visa accepted.

PORTO, Portugal

Pedras Rubras Airport (Francisco Sá Carneiro), 8 mi (13 km) N

Esc1500-1700 ($10-11.40), 15-25 min. Tip 5%.

No. 56 from outside arrival hall every 15-30 min 6 am-11:38 pm. Esc12 (85¢). Baggage space. Many stops into town (Boavista Sq., Galiza Sq., Carmo). 30-45 min ride.

Avis, Budget, Europcar, InterRent.

P Esc70 (50¢)/hour.

PRAGUE, Czechoslovakia

Ruzyne Airport, 11 mi (17 km) W

K200 ($12), 30-60 min depending on traffic. Small surcharge at night. Tip: Round up meter, adding 10%.

Czechoslovak Airline bus meets flights. Pay K8 (50¢) aboard. Limited baggage space but comfortable 30-40 min ride to Vltava Terminal in Revolucni S

No. 119 every 10-20 min from outside arrival terminal. Fare K4 (10¢). 35-60 min into town. Dejevicka St. (Prague 6) stop near the metro station is convenient to many hotels. Very limited baggage space. Service 6 am-11 pm. Abbreviated weekend, holiday schedule.

Pragocar represents Avis, Budget, Denzel, Godfrey Davis/Europcar, Hertz, InterRent.

PRESTWICK (Glasgow), Scotland

Prestwick Airport, 29 mi (48 km) S of Glasgow

£23 ($38) into city, no extras. Tip 15%. 45-55 min.

No. 501 via Glasgow Buchanan Bus Station to Edinburgh St. Andrew Sq. Bus Station. Mon-Sat hourly 8:38 am-5:38 pm. Sun every 2 hrs 10:08 am-8:08 pm. Boards immediately outside terminal. 70 min to Glasgow, 150 min to Edinburgh. £2 ($3.30) to Glasgow, £4 ($6.60) to Edinburgh.

Railair Link shuttle bus to Prestwick Rail Station meets arrivals, departures. 5-min trip. From Prestwick station, trains every 30 min to Glasgow Central Station. First train 6:20 am, last 11:05 pm. Sunday hourly service 9:54 am-11:04 pm. 45-min trip. Good baggage space. £2.75 ($5).

Avis, Europcar, Hertz, Swan National Eurodollar.

P £3 ($5)/day 1-6 days; £2.20 ($4)/day 7+ days.

PROVIDENCE, Rhode Island

Theodore Francis Green State Airport, 7 mi S

$16-17, 15-20 min.

Airport Limousine at 60-75 min intervals 5:25 am-12:55 am. $5.75.

Avis, Budget, Hertz, National, Snappy, Thrifty.

P ST $20, LT $5/day, $18/week.

PUERTO VALLARTA, Mexico

Gustavo Diaz Ordaz Airport, 4 mi (6 km) N

$3, 10-15 min.

Colectivos available in front of terminal 8 am-last flight. Drop-off anywhere along route into city. $1.

Avis, Budget, Dollar, Hertz, National.

QUEBEC, Quebec
Aeroport de Quebec, 10 mi NW

C$20-22 ($17.60-19.35), 25-30 min.

To Quebec, 25 min., stops at major hotels, OW C$7.50 ($6.60), RT C$13 ($11.45); to Ste. Foy OW C$5.50 ($4.85), RT C$9 ($7.90). Frequent departures M-F 9:05 am-10 pm, Sat 10:45 am, 4:15 and 6:55 pm. Sun 10:45 am, 1:05, 4:15, 6:55, 8:55, and 10:45 pm. Return schedule posted in hotel lobbies.

Budget, Hertz, Thrifty, Tilden.

P ST C$8.60 ($7.55), LT C$5.35 ($4.70)/day.

QUEENSTOWN, New Zealand
Frankton Airport, 4.4 mi (7 km) E

NZ$12 ($7.95), 10 min. Call Southern Taxis 27-888.

Prestige Airport Shuttle meets flights. NZ$4 ($2.65). 15 min.

Avis, Budget, Hertz.

RALEIGH/DURHAM, North Carolina
Raleigh-Durham Airport, 14 mi NW of Raleigh, 6 mi SE of Durham

To Raleigh $20-21, to Durham $20-22.

Raleigh Transportation, 6:30 am-11:30 pm. $14 first person, $3 each addl to major hotels, campuses. Res & Info: 919-821-2111. AMEX, MC, VISA.

Avis, Budget, Dollar, Enterprise, Hertz, National, Snappy, Thrifty.

P ST $6, LT $3.50.

REGINA, Saskatchewan
Regina Airport, 3 mi (5 km) SW

C$7 ($6), 10 min. Tip 10%. If no cabs at airport, use courtesy phone.

Avis, Budget, Hertz, Thrifty.

P C$6 ($5.10)/day.

RENO, Nevada
Reno Intl Airport, 3 mi SE

$8, 7-10 min.

Airport Limousine, $2.25. 10-15 min to hotels in Reno, Sparks.

From front of terminal. 75¢. 20 min to City Center Transfer.

Avis, Budget, Dollar, General, Hertz, National.

P ST $10, LT $4.50.

REYKJAVIK, Iceland

Keflavik Airport, 31 mi (50 km) SW

Bus meets flight. Kr350 ($8), children Kr200 ($5). 45 min to hotels Loftleidir, Esja.

Arnarflug, Avis, Hertz.

RICHMOND, Virginia

Richmond Intl Airport, 7 mi SE

$16-30, addl passengers 30¢ each. 20-30 min.

Groome Transportation meets flights. Into town: $11.50 one person; $16.50 for two; $19.75 three. MC, VISA, AMEX accepted.

Alamo, Avis, Budget, Dollar, Hertz, National, Payless, Snappy, Thrifty.

P ST $8, LT $5.

RIO de JANEIRO, Brazil

Galeao Airport, 12.5 mi (19.3 km) NE

To city center, Cr3120 ($8–U.S. currency accepted). 25-40 min to Inter-Continental Hotel. Tip 10%. Brazilian officials advise tourists to ignore counter marked "R.D.E." offering assistance with taxis and "passports." Look just beyond for "Rio de Janeiro State Tourism Authority," which sells taxi vouchers at standard rates and offers other legitimate services.

Airport bus with baggage space, air conditioned (hence called the "Frescao" in Portuguese). Fare to driver Cr660 ($1.70). Departs hourly. Stops at Santos Dumont, the downtown airport, and all beaches and major hotels, terminating at Hotel Nacional in about 1 hr.

Avis, Budget, Hertz, Localiza, Rentarauto.

P Cr1000 ($2.40)/day.

RIYADH, Saudi Arabia

King Khaled Intl, 17.5 mi (30 km) N

SR35-75 ($9.50-$20.40), 35-45 min. Flat rates depend on distance—pay airport cashier in advance. Travelers advised to hire only airport-authorized taxis, identifiable by logo on side of vehicle.

SAPTCO bus is a luxurious intercity model. Fare SR5 ($1.35). Departures every hour 6 am-11 pm. 50-60 min ride. Stops at Al Batha and some major hotels.

Avis, Budget, InterRent.

ROANOKE, Virginia
Roanoke Regional Airport, 4 mi NW

$10, addl passengers $1 each. 10 min.

Limousine Services: Blacksburg, C'artier, Roanoke Airport Limo.

Avis, Dollar, Hertz, National.

ROCHESTER, Minnesota
Rochester Municipal Airport, 8 mi S

$14-15, 15 min.

Avis, Budget, Hertz, National.

ROCHESTER, New York
Monroe County Airport, 5 mi SW

$8-10, addl passengers $2 each. 10-20 min.

No. 2 Thurston-Parsells bus at 15-30 min intervals 5:50 am-6:33 pm M-F only. 80¢ exact. 28 min to Main & Clinton, downtown.

Ajax, American Intl, Avis, Budget, Dollar, Hertz, National, Snappy, Thrifty.

ROME, Italy
Leonardo da Vinci Airport (Fiumicino), 18 mi (30 km) SW

Lit50,300 ($40.25), 45-60 min to central Rome. Lit500 (40¢) per bag. Fiumicino Guide says, "We strongly advise you to use only licensed taxis with meters and to decline any other offer of transportation." To Rome, fare is meter plus Lit10,000; from Rome, meter plus Lit14,000. Flat fare, no tip.

Acotral bus every 15 min 7 am-12:45 am, then hourly 1:30-6:30 am. Buy ticket for Lit5,000 ($4) at Acotral office outside arrivals area. Plenty of luggage space. Trip to Via Giolitti, at side of Rome RR station, about 55 min.

Avis, Eurodrive, Europcar, Hertz, InterRent/Autotravel, Italy by Car, Maggiore.

P ST Lit24,000 ($19.20), LT Lit10,100 ($8).

ROTTERDAM, Netherlands
Rotterdam Airport, 5 mi (8 km) N

Dfl25-30 ($13-16), 15 min. The Hague, Dfl55 ($29.15), 25 min.

No. 33 to Central Station about every 6-15 min. Local service, 25-min ride. Dfl2.50 ($1.30).

Avis, Budget, Europcar/InterRent, Hertz.

P ST Dfl10 ($5.30), ST Dfl5 ($2.65).

SABA, Windwardside, Netherland Antilles
Saba Airport, 3 mi (5 km) N

$6.25, 5 min; St. John's, $8.75, 20 min; the Bottom, $10, 25 min; Fort Bay, $13, 30 min.

Doc's, Johnson's, Har Diana.

SACRAMENTO, California
Metropolitan Airport, 11 mi NW

$25-30, 15-20 min. "Taxi and limousine service has been deregulated. Travellers are advised to confirm costs," airport advises.

Skyline Airporter, Downtown Airporter, $7.25-9 to downtown hotels, 15-20 min. **To Chico, Yuba City, Oroville, Marysville, Paradise** Airport Transportation Service. Res: 800-832-4223. **To Wheatland, Live Oak, Sutter, Beale AFB** Yuba City Airporter Service. Res: 916-671-1199.

Door-to-door Sacramento city and county: Gold Dust Airport Shuttle, 916-944-4444; Sacramento Metro Shuttle, 916-962-1222; and others.

Avis, Budget, Dollar, Hertz, National.

P ST $9, LT $4.50.

SAGINAW/BAY CITY/MIDLAND, Michigan
Tri City Airport, 10 mi NW of Saginaw, 11 mi SE of Midland, 13 mi SW of Bay City

Available.

Avis, Budget, Hertz, National, Thrifty.

SALT LAKE CITY, Utah

Salt Lake City Intl Airport, 3 mi (5km) W

$11, 10-15 min downtown. To Park City $50, Ogden $60, Provo $65, Alta $43, Logan $110.

To Salt Lake City Hourly or more frequently 6:31 am-11:50 pm; greatly reduced service on Sunday. **To Park City** Park City Transportation Service, at noon, 3, 8, 10 pm. $13. **To Ogden, Logan, Hill AFB, Morgan, Roy, Brigham City, Tremonton** Key-North Limousine at 6:30, 8:30, 10:30 am; 1, 4:30, 7:30, 9:30, 11 pm. Sat-Hol 8:30, 11 am; 5, 10:30 pm. $16 to Ogden; $25 to Logan, minimum 4 passengers. **To Orem, Provo & South** Key Limousine Service at 9, 11 am; 1, 3, 5:30, 7:30, 9:30, 11 pm. $16 OW, $27 RT. Info & Res: 800-678-2360.

To ski resorts Scheduled service available during season. Inquire at ground transportation counter.

No. 50/150 Airport/Intl Center bus every 30-60 min 6:31 am-11:45 pm M-F. Sat hourly 6:35 am-6:35 pm. Fare 50¢. Limited Sun service, no Hol. Bus stops between terminals 1 & 2. 30-min ride to 355 S. Main St., downtown.

Alamo, Avis, Budget, Dollar, General, Hertz, National, Thrifty.

P ST $5, LT $3/day, $15/week. Free shuttle every 10 min.

SALZBURG, Austria

Salzburg Airport, 4.25 mi (7 km) SW

S80-90 ($6.70-7.50) for 3 persons, S5 (40¢) for the fourth. To downtown hotels, 10-15 min. Tip 10%.

No. 77 every 20 min 5:54 am-11:11 pm to main central station. S12 ($1), buy ticket on bus. 15 min to Bahnhof. Sunday service less frequent. No room for large baggage.

Avis, Budget, Denzel, Hertz, InterRent.

P S80 ($6.70)/day dropping to S60 ($5)/day after 8 days.

SAN ANTONIO, Texas

San Antonio Intl Airport, 8 mi N

$12, 15-20 min.

SuperVan to downtown hotels, $7 OW, $12 RT. 20 min. Reservation required for return: 512-344-7433.

No. 12 Airport Express M-F only. Operates to the airport from Commerce & Alamo, downtown, approx every 50 min 6:38 am-6:25 pm. 20-30 min ride downtown. 75¢. Bus stop at VIA sign next to taxi stand.

Advantage, Alamo, Avis, Budget, Dollar, Enterprise, General, Hertz, National, Payless, Snappy, Thrifty.

P ST $6, LT $5, $3.50.

SAN DIEGO, California

San Diego Intl Airport, 3 mi NW

$7-9. 10-15 min downtown. Hotel Circle, $14-18, 15-20 min. Mission Bay hotels, $13-17, 10-15 min.

Door-to-door shared-van service to downtown, Coronado, Mission Hills, Normal Heights, Point Loma, Chula Vista, La Mesa, San Carlos, Rancho Bernardo, Escondido, La Jolla, Oceanside, El Cajon, San Ysidro. Operators: Peerless, 619-695-1766; Coast Shuttle, 619-477-3333; Cal Pac Transport, 619-485-0202; Greoware Limo, 619-589-7885; SuperShuttle, 619-278-8877. **To Oceanside, Camp Pendleton** West Wind Express daily except Mon-Tues at 1:30, 4:40, 7:30, 10:30 pm. $12. On call other times: 619-670-3232.

No. 2 bus every 10 min M-F 5:20 am-11:36 pm; every 30 min Sat-Sun-Hol 6 am-8:29 pm, then every 45 min to 12:14 am. Fare $1.25. 15-20 min to 3rd & Broadway, downtown.

Airways, Alamo, Avis, Budget, Dollar, General, Hertz, National, Payless, Thrifty, Tropical.

P ST $12 first day, $18 each day thereafter. LT $10.

SAN FRANCISCO, California

San Francisco Intl Airport, 14 mi S

$29 downtown, Fisherman's Wharf area $34. Up to five can share a ride for flat $24. Limit of three drop-offs. Nonstop downtown run is 20 min.

SFO Airporter to Meridien, Grand Hyatt, Westin St. Francis, Parc 55, Marriott, Nikko Hilton, Ramada Renaissance, Bart/Muni connection. Every 20 min 6:10 am-11:10 pm. $7 OW, $11 RT. Board lower level center section. Info: 415-673-2433.

Door-to-door San Francisco & vicinity Numerous shared-ride vans on demand from upper level center island. Fares $8-11 within city, $8-36 San Mateo County; $13-42 Contra Costa County; $8-78 Santa Clara County. Res & info: SuperShuttle 415-558-8500; Door to Door Airport Express

415-775-5121; Lorrie's 415-334-9000; Yellow 415-282-7433; South Bay Airport Shuttle, 800-548-4664; Bayporter Express 415-467-1800 (CA); Direct Shuttle 415-674-0474; Francisco's Adventure 415-821-0903.

SamTrans buses No. 7B, 7F daily 5:43 am-1:21 am at 15-, 30-min intervals. No luggage on 7F. From upper level at SamTrans sign. 7B $1, 7F $1.25 exact. 30-50 min ride to Transbay Terminal, 1st & Mission, downtown.

No. 3B bus from SamTrans boarding area upper level, 50¢ to Daly City BART station. Bus operates 5:52 am-6:22 pm every 30 min M-F, every 30-60 min 8:30 am-4:30 pm Sat-Sun-Hol. BART cost $1.10 addl to downtown. Scenic but no fun if you are wrestling luggage.

To Oakland Intl Airport Bayporter Express on the hour 6 am-11 pm with stops on call at Treasure Island, Oakland Army Base, Parc Oakland. $10, 60 min. **To San Jose Municipal Airport** Greyhound. Info: 415-877-0366.

Able, Alamo, American Intl, Avis, Avon, Budget, Dollar, General, Go Vacations, Hertz, Lindo's, National, Pacific, Payless, RPM, Showcase, Snappy, Thrifty, U.S. Premier.

P ST $15, LT $9. Free shuttle every 5-15 min. Valet $25.

To Marin County Marin Airporter every 30 min 5:30 am-Midnight. Sausalito, Mill Valley $9; Larkspur $10; Terra Linda $11; Ignacio $12; Novato $13. Santa Rosa Airporter hourly 6:35 am-11:35 pm to Corte Madera Inn $12; Mill Valley $10; Novato $14; Santa Rosa $17; Petaluma $17. Info: 415-898-8888. **Concord/Berkeley** Airport Connection to Emeryville, Durant, Claremont, Orinda, Walnut Creek, Concord. Info: (US) 800-AIRPORT, 415-363-1500. **To Travis AFB, Vallejo** Travis/Solano Airporter, 707-437-4611. **To Fremont, Union City, Newark** Fun Connection every 2 hrs 7 am-10 pm. $13. 415-791-7160. **To Santa Rosa, Rohnert Park, Petaluma** Sonoma County Airport Express hourly 6 am-midnight. $14. **To Sonoma, Boyes Spring, Kenwood, Glen Ellen** Sonoma Airporter at 8:05, 9:45 am; 12:20, 2:45, 5:40 pm weekdays; 8:05 am, 12:20 pm, 5:40 pm, weekends. $20-25. **To Napa, Vallejo** Evans Airport Service about every 1 hr 45 min 6:30 am to 10:45 pm. Vallejo $14, Napa $15. **To Castro Valley, Dublin, San Ramon, Danville, Pleasanton, Alamo** San Ramon Valley Airporter Express about every 90 min 6:30 am-11:30 pm. $19.

All suburban services board center island, lower level.

SAN JOSE, California

San Jose Intl Airport, 3 mi NW

$9.50-10, 7-10 min. Same fare and travel time to bus, RR stations. To Sunnyvale $11-18; Cupertino $19-22; Palo Alto $24-30, Stanford U $30.

No scheduled service downtown. **To Santa Cruz** Peerless Stages has

7 departures 8:35 am-8:25 pm. Intermediate stops at Los Gatos, Scotts Valley. $4.05 to Santa Cruz. Several companies provide shared door-to-door service to **South Bay** areas. Typical fares: San Jose $12-15; Santa Clara $12-14; San Francisco Airport $16-24; Palo Alto $12-15. Some fares cover two people– confirm. SuperShuttle, 408-629-8500; South Bay Airport Shuttle, 408-559-9477; BayPorter Express, 415-467-1800.

Santa Clara County Transit No. 10 from bus shelter outside baggage claim area. To downtown, Greyhound, and train station. Departures every 15-30 min 5:29 am-12:11 am. To Fremont, Mipitas, or BART, transfer at Civic Center to express No. 180. Connections for entire Santa Clara County at Civic Center or downtown. Fares $1. Free connection to Metro Plaza Light Rail station aboard Metro Shuttle, every 10 min 6 am-8:30 am and 4 pm-7 pm M-F. Available mid-day by calling 408-452-0111.

To San Francisco Intl Airport (SFO) BayPorter Express. $16. Reservation required: 415-467-1800.

Able, Avis, Avon, Budget, Dollar, General, Hertz, National, RPM, Thrifty.

P ST $16, LT $6. Shuttle every 5 min to LT lot.

SAN JOSE, Costa Rica

Juan Santa Maria Airport, 10 mi (16 km) NW

C800 ($9), 20 min. To Puntarenas Port, C4000 ($45), 1 hr 30 min; to Limon Port, C4500 ($50), 1 hr 45 min; to San Carlos, C5000 ($55), 2 hrs; to San Isidro del General, C6000 ($67.50), 3 hrs; to Liberia, C5000 ($55), 3 hrs. Cabs are orange colored. No tips.

Microbus for 5-10 persons, C1000 ($11) total to San Jose.

Alajuela-San Jose bus every 20 min 4 am-11 pm from stop across street from airport. Fare C30 (35¢). Baggage OK.

Ada, Avis, Budget, Dollar, Hertz, National, Nikel, Santos, Toyota.

P C200 ($2.25)/day, C2000 ($22.15)/month.

SAN JUAN, Puerto Rico

Puerto Rico Intl Airport, 5 mi E

$10, 15 min.

$2.50, 20 min.

Avis, Budget, Hertz, National, Thrifty.

SAN PEDRO SULA, Honduras

LaMesa Airport, 9 mi (15 km) E

£15 ($5.40), 15 min. Haggling over price not uncommon. A good bargainer may get the price down to £10 ($3.60).

Blitz, Budget, Molinari.

SAN SALVADOR, El Salvador

El Salvador Intl Airport, 30 mi (48 km) SE

CO80 ($16), 25 min to city center, hotels. Tip 10%.

ACACYA coach at 11 am, 5, 7 pm daily. CO15 ($5), 30 min.

Avis, Budget, Hertz, Imosa, Superior.

P CO9 ($1.80)/day.

SANTA BARBARA, California

Santa Barbara Municipal Airport, 6 mi W

$16, 15 min.

Airbus Express $10 downtown. Also door-to-door service other areas. Info: 805-964-7759.

MTD No. 11 bus from street outside airport every 20-30 min 6:05 am-11:13 pm. To Goleta, UCSB, downtown Santa Barbara. 75¢.

Avis, Budget, Dollar, Hertz, National, Thrifty.

SANTA FE, New Mexico

Santa Fe Municipal Airport, 9 mi SW

$13 flat fare plus tax, $1 per person for more than 3. 15-20 min.

To/from Albuquerque Shuttlejack Coach from Inn at Loretto and Hilton Inn (5 min later) to Albuquerque at 5, 7, 9, 9:45, 11 am; 1:15, 1:55, 3:30, 5:35, 8:30 pm. From Albuquerque at 6:50, 8:55, 11:10, 11:55 am; 2, 3:25, 5:30, 6:45, 8:15, 10:15 pm. 75-min trip. $20 OW. Res: 505-982-4311.

Avis, Budget, Hertz.

SANTIAGO, Chile

Santiago Intl Airport, 10 mi (16 km)

$10-12, 30 min to city center. No tip.

Every 30 min 6:30 am-9 pm to Moneda 1523, city center. Fare 90¢. 30-min ride. Look for bus in front of airport.

Atal, Avis, Budget, Chilean, Dollar, Hertz, National, Rentauto.

SANTO DOMINGO, Dominican Republic

Aeropuerto Intl de las Americas, 17 mi (27 km) W

DRP$35 ($13), 25 min. No tip necessary.

Avis, Budget, Cima, Cumbre, Dollar, Express, Hertz, Nelly, Patsy, Puerto Rico.

SAO PAULO, Brazil

Congonhas Airport, 9 mi (14.4km) SW

Red/white Radio cabs, Cr$7.000 ($15), 15 min. Also small cabs, carry two people and a few pieces of baggage. Look for cab dispatcher.

Bus to city center every 5 min on Ave. Washington Luis, in front of airport. Cr20, 35-40 min. No room for baggage, often crowded and uncomfortable.

To Sao Paulo Intl Airport (Guarulhos) Taxi Cr$16.550 ($37). Also Metro bus every 30-45 min 5:45 am-9:25 pm. Cr$3.500 ($8).

Avis, Hertz, Interlocadora, Localiza, Locarauto, National, Unidas.

P Cr$2.550 ($5)/day.

SAO PAULO, Brazil

Sao Paulo Intl Airport (Guarulhos), 18 mi (30km) NE

Blue/white Radio Taxis. Cr$10.100 ($22), 25-30 min to Republica Sq. Purchase taxi ticket from agent in terminal.

Metro bus every 25 min 5 am-Midnight, then hourly. Cr$3.500 ($8). Purchase ticket in advance on ground floor in national arrivals section. 35-40 min to Praca de Republica (Republic Sq).

To Congonhas Airport for Rio Shuttle, domestic flights Taxi Cr$13.600 ($30). Also Metro bus hourly 6 am-10 pm. Cr$3.500 ($8).

Avis, Hertz, Interlocadora, Localiza, Locarauto, Nobre, Unidas.

P Cr$2.550 ($5)/day.

SAO PAULO, Brazil

Viracopos Airport, 56 mi (90 km) W

Cr$40.000 ($89), 90 min. To Sao Paulo Intl Airport (Guarulhos), Cr$46.800 ($105), 80-100 min. Tip 10%. Taxis are not metered. Because of the great distance to Sao Paulo, driver can charge round-trip rate for a one-way ride. Best to agree in advance what the fare will be.

Airlines provide free transport between this charter-flight airport and the city. Inquire.

Localiza.

P Cr$8.000 ($17)/day.

SARASOTA/BRADENTON, Florida

Sarasota-Bradenton Airport, 6 mi N of Sarasota

$8, 15 min to Sarasota; $14, 25 min to Bradenton.

Ajax, Alamo, American Intl, Avis, Budget, Dollar, Enterprise, Hertz, National, Payless, Snappy, Thrifty.

P ST $7.50, LT $4.

SASKATOON, Saskatchewan

Saskatoon Airport, 4.5 mi NW

C$8.50 ($7.50), 15-20 min. Tip 10%.

Budget, Hertz, Thrifty, Tilden.

P $6/day.

SAVANNAH, Georgia

Savannah Municipal Airport, 11 mi (17.5 km) NW

$15 flat fare, $3 each addl passenger. 15-20 min.

Agency, Alamo, Avis, Budget, Dollar, Enterprise, General, Hertz, National, Snappy, Thrifty.

SEATTLE/TACOMA, Washington

Seattle-Tacoma Intl Airport, 13 mi S of Seattle, 23 mi N of Tacoma

Seattle center, ferry dock, hotels, $22.50-28, 20-40 min; U. of Washington, $26-30; Renton $13-16; Bellevue $25-27, 30-45 min; Kirkland, $27-37; Federal Way, $17-20; Tacoma, $32-40, 30-45 min.

Gray Line Airport Express to downtown Seattle every 15-30 min 6:15 am-11:45 pm. $7 OW, $12 RT. 25-50 min. Stops at Stouffer, Crowne Plaza, Four

Seasons, Hilton, Sheraton, Westin, Warwick, 8th & Bell Hotels, Best Western. Not all stops each run. Ask agent. Info: 206-626-6088. **To Tacoma, Olympia** Capital Aeroporter every 90 min. Also serves Shelton, Centralia, Chehalis, Hoodsport, Tumwater, Lacey, Puyallup, Lakewood, Fife, Federal Way. Info, res: 206-754-7113 **To Bellingham** Airporter 8, 11 am; 1, 3, 5, 7 pm. $21 OW, $38 RT. **To Vancouver, B.C.** Quick Coach. 12:30, 3, 10 pm; $32. **To Everett, Lynwood** Everett Airporter hourly 6:30 am-12:30 am, $7-11. **To Ft. Lewis & McChord AFB** Ft. Lewis/McChord Airporter at 7 am then every 2 hrs to 11 pm. $7-8 **To Bellevue & N of Lake Washington** Suburban Airporter every 30-60 min 5:35 am-12:30 am. 25 min, $9.50 downtown.

ShuttleExpress door-to-door, shared-ride van service on demand. Downtown Seattle $14-16, Bellevue $11-13.50, Everett $12-13, Kirkland $12.50, Mercer Island $16, Federal Way $16, McChord AFB $17, Redmond $16-20, Renton $10-12, Bothell/Woodinville $16-20. Info: Courtesy phone No. 48 or 206-622-1424. In Washington 800-942-0711. Also, "Standhail" limos on premises: Seattle $27, Bellevue $30, Tacoma $35. Fare covers all passengers. Inquire at info desk.

No. 174/194 bus every 30 min 5 am-12:40 am. Hourly service Sun from 7:05 am. $1.25 peak times, 85¢ off peak. Buses depart S end of baggage claim. Look for Metro sign. 40 min run to 4th Ave. & Union St. downtown. **To Renton, Bellevue, Aurora Village** No. 340 bus about every 30 min 5:59 am-9:15 pm. Hourly 8:22 am-9:22 pm Sat, Sun. **To Tacoma** No. 174 bus southbound. Transfer at Federal Way to Pierce No. 500 for downtown Tacoma.

Alamo, Avis, Budget, Dollar, General, Hertz, National, Sears, Thrifty.

P First 30 min free. $10/day. VISA, MC. No LT lot on airport grounds but private lots available at $6/day; free transportation.

SEOUL, Korea

Kimpo Intl Airport, 11.5 mi (19km) NW

Regular cabs OK for 3 adults plus bags. For 4-5 psgrs, a "Call" cab required, double fare. Sample regular fares: W6000-8000 ($8.50-11). Rates to all major hotels posted at loading zone. No tip.

Two routes, same fare of W500-1500 (70¢-$2.10) and 10-min frequency on each. **No. 600** To Sports Complex, Jamsil via Heuk Suk Dong, Seoul Palace Hotel, Express Bus Terminal, Riverside Hotel, Youngs-dong, Nan Seoul Hotel, Aid Apt, Koex. **No.601** To Sheraton Walker Hill Hotel via City Hall Plaza hotels: President, Lotte, Seoul Plaza, Westin Chosun. About 30-, 40-min ride to town. Ample baggage space, very comfortable. Boards at center, arrivals hall.

Three express bus routes, OK for commuters with light or no luggage. No. 63 to City Hall. No. 68 to railway station, City Hall area. No. 700 to Yongdungpo-Yoido. Departures every 10 min 6 am-10 pm. Fare W400 (60¢).

Hertz, Korea, Sam-Bo, and others.

P W1000 ($1.40)/day.

Note: Seoul subway, though it does not connect to airport, is fast, clean, safe, well marked in English, and very inexpensive.

SHANGHAI, China

Hongqiao Airport, 9.3 mi (15 km) W

RY15.60 ($3.30), 25 min. No tip.

CAAC provides transfer from airport to city ticket office. RY1.8 (40¢), 30 min.

Self-drive cars not available in China, but chauffeur-driven are. Inquire at hotel.

SHANNON, Ireland

Shannon Intl Airport, 15 mi (24 km) NW of Limerick

£14 ($22.10), 30 min into town. Tip optional.

Bus Eireann to Limerick RR station about every 60 min from 7:05 am, less frequent Sat. Sun service. Last evening bus 11:15 pm. Baggage room. Pay £3-4 ($4.75-6.30) fare aboard. About 45-min ride.

Avis, Budget, Bunratty, Eurodollar, Europcar, Hertz, Irish, Kenning.

SHREVEPORT, Louisiana

Shreveport Regional Airport, 5 mi SW

$8.25, 15 min.

Avis, Budget, Hertz, National, Thrifty.

SINGAPORE

Changi Airport, 12.5 mi (20 km) NE

S$13 ($7.65) including airport surcharge of S$3 ($1.75), 25 min. Surcharge midnight-6 am, 50% of metered fare.

No. 390 from Basement 2 of terminal every 10-12 min 6 am-midnight. Pay driver S90¢ (53¢) exact change. No baggage racks. 40-50 min to city center. Clean and reasonably comfortable.

Avis, Sintat.

P S$10 ($5.90).

SIOUX FALLS, South Dakota
Joe Foss Field, 3 mi N

$4.50, 10 min.

Avis, Budget, Dollar, Hertz, National.

SOUTHEND, England
Southend Airport, 2 mi (3.2 km) N of Southend, 35 mi (56 km) E of London

£2 ($3.50) to Southend, 10 min. Cab to London—£35 ($60), 60-90 min—not recommended.

Hail and ride shuttle bus between Central Station, Victoria Station and airport. 70p ($1.20).

Bus or taxi to Rochford Station, 6 min. Then British Rail connection to London Liverpool St. Station. Three trains per hour, 58-min ride. £7 ($12.15).

Godfrey Davis/Europcar.

P ST £3 ($5.20), LT £1.10 ($1.90).

SPARTANBURG, South Carolina See GREENVILLE/SPARTANBURG, South Carolina

SPOKANE, Washington
Spokane Intl Airport, 6 mi SW

$12, addl passengers 50¢ each. 15-20 min.

Avis, Budget, Dollar, Hertz, National, Thrifty.

SPRINGFIELD, Illinois
Capital Airport, 3 mi N

To state office complex, one person, $5; two or more, $4 each; 10-15 min. To Holiday Inn East Conference Center, $10 per person, 2 or more, $7 each; 20 min. To Sangamon State U., $12 per person, 2 or more, $9 each; 25-30 min.

Avis, Budget, Hertz, National.

SPRINGFIELD, Massachusetts

See HARTFORD, Connecticut/SPRINGFIELD, Massachusetts

SPRINGFIELD, Missouri

Springfield Regional Airport, 8 mi NW

$7.65, addl passengers 50¢ each. 15 min.

Avis, Budget, Hertz, National.

ST. CROIX, Virgin Islands

Alexander Hamilton Airport, 7.5 mi SW

$10 for 1-2 persons, $5 each addl. 20 min to Christiansted; to Frederiksted, $8 per person, 20 min; to Queens Quarter Hotel, $8 per person, 10 min. Baggage: first piece free, second 50¢. 75¢ for trunk.

Avis, Budget, Hertz, Maurice.

ST. LOUIS, Missouri

Lambert St. Louis Intl Airport, 13 mi NW

$18 downtown, 20-30 min. Addl passengers 50¢ each. To Clayton $13.

Airport Express every 20 min 6:30 am-10:30 pm. $7 OW, $10 RT to downtown, midtown, Clayton hotels. Info: 314-429-4940.

Avis, Budget, Dollar, Hertz, National.

P ST $8, LT $6/$4.

ST. LUCIA, West Indies

Hewanorra Intl Airport, 33 mi SE of Castries; Vigie Field, 2 mi N of Castries

From Hewanorra $30, 60 min. From Vigie $4, 5 min. Vigie to Cariblue Hotel $10; St. Lucian $8; Halcyon Beach $4.

Avis, National.

ST. MAARTEN, Netherlands Antilles

Princess Juliana Intl Airport, 7 mi W

To Philipsburg, $8 for 1-2 passengers, $2 each addl. 20 min. To Marigot (French capital), $8, 15 min; to Orient Beach, $20, 40 min; to Grand Case, $16, 40 min.

Avis, Beach Island, Budget, Cannegie, Caribbean, Diamond, Dollar, Hertz, Holiday, National, Opel, Reggie's, Reynolds, Risdon, Roy Rogers, Speedy, Sunshine.

P 50¢/hour.

ST. PAUL, Minnesota See MINNEAPOLIS/ST. PAUL, Minnesota.

ST. PETERSBURG, Russia

Leningrad Pulkovo Airport, 10.5 mi (17km) S

Rb5-7 ($8.25-11.55 at official exchange rate), 15-20 min. Tip: Rb1.

Aeroflot service to office on Nevsky Prospekt.

Express bus links airport and Pobyedi station of Metro. 20-min run. Metro fare 5 kopecks (10¢) through turnstile. Nevsky Prospekt, center city, is sixth stop. Total travel time 40-60 min including connections. Cabs available on street. **Note:** Although most Westerners rely on Intourist to make all travel arrangements in the Soviet Union, it is quite possible to do so oneself in the major cities. Traveling unescorted on the Metro is a superb introduction to everyday life. Ask Intourist for English language maps; carry kopecks for fares.

Intourist makes arrangements.

ST. THOMAS, Virgin Islands

Cyril E. King Airport, 3 mi W

$4.50 first psgr, $4 each addl, 10 min to Charlotte Amalie.

Avis, Budget, Hertz, V.I. Auto Rental.

STAVANGER, Norway

Stavanger Airport, 9 mi (14.5 km) SW

NKr160 ($24), add 25% at night. NKr2-3 each addl passenger, NKr1 each bag. 20 min to town. Tip not essential but up to 10% for good service.

SAS Airport Bus from stop opposite taxi rank. NKr28 ($4.20), 20 min to main hotels—Atlantic, SAS, Victoria. No luggage racks but wide aisles. Departures every 30 min except no departures Sun until 4:45 pm.

SAS Airport Bus, NKr28 ($4.20). M-F regular departures 7:50 am to 11:45 pm, Sundays 2:55 pm to 11:50 pm. 20-30 min to Atlantic, SAS, Victoria hotels. No luggage racks but wide aisles.

Avis, Budget, Europcar, Hertz.

P NKr35 ($5.25)/day.

STOCKHOLM, Sweden
Arlanda Airport, 26.25 mi (42 km) N

SKr 270-330 ($44-54), 35-40 min. Tip 10%. A less costly weekday alternative between 8 am and 4 pm: look for "Arlanda Retur" car. Fare Skr 150 ($24.50).

SAS car (may be shared ride): SKr 185-275 ($30-45) per person into city. Return car must be ordered at least 4 hrs in advance, phone 797-3700.

Arlanda Airport (SL) bus to Central Station (Vasagatan), Stockholm City. 7:10 am-10:30 pm at 10-,20-min intervals; thereafter meets flights. SKr35 ($5.70). Pay driver or purchase ticket in advance at airport or city terminals. Baggage racks. 40-45 min. All buses stop on north side of Stockholm at Haga Park Air Terminal and Ulriksdal Air Terminal. Central Station is at Vasagatan 6-14. To Brommaplan, Sundbybergs Torg, Kista Centrum: Departures every 30-60 min 6:20 am-11 pm. **To Uppsala** Bus No. 801 every 30 min 5:50 am-12:20 am. SKr22 ($3.60). 40-min journey.

Avis, Budget, Europcar, Hertz, InterRent.

P SKr 35 ($5.70)/day.

STUTTGART, Germany
Stuttgart/Echterdingen Airport, 9 mi (14 km) S

DM30 ($17.40), 15-20 min into town. Stuttgart is birthplace of the Mercedes Benz and there are plenty of them waiting at the cab stand.

SSB Express bus at 5:25, 6:25, 7:25 am, then every 20-40 min to 12:15 am. Baggage room, comfortable seats. 20 min to Central Air Terminal and Hauptbahnhof. Fare DM6 ($3.50).

Lufthansa Airport Express It "flies" 2 times daily non-stop to Frankfurt–Main Airport, 2 hrs.

Autohansa, Avis, Europcar, Hertz, InterRent, Mages, Scheer, Sixt-Budget.

P ST DM30 ($17.40), LT DM5 ($2.90).

SUN VALLEY, Idaho
Friedman Memorial Airport, 16 mi (26km) S

A-1 Taxi, $13, 20 min.

Elkhorn Lodge, Sun Valley Lodge, Warm Springs Lodge operate courtesy limos for registered guests.

Avis, Budget, Hertz, National, U-Save.

P $3/day.

SURFERS PARADISE, Australia
Coolangatta Airport, 15.5 mi (25 km) S

A$20 ($15.80), 35 min.

EET coaches meet flights. A$5 ($4), 30-45 min to major hotels on the Gold Coast. Buy ticket in terminal.

Avis, Budget, Hertz, Thrifty.

P A$10 ($7.90)/day.

SYDNEY, Australia
Kingsford Smith Airport, 6.2 mi (10 km) S

A$12-15 ($9.50-11.85), 20-30 min to Regent Hotel, city center, Kings Cross.

No. 300 Airport Express every 20-30 min 6:25 am-10:15 pm from intl terminal, stopping at domestic terminal, then to Central Station and Circular Quay downtown, near hotels. A$3 ($2.30), children half fare.

Kingsford Smith Transport bus to all major hotels in city and King's Cross every 30 min 7 am-5 pm. To city A$3.40 ($2.60), children half fare.

To Wollongong Watts Bus Service, A$11 ($8.50).

Avis, Budget, Hertz, Thrifty.

P ST A$11 ($8.50); LT A$6 first day, thereafter A$5 ($3.80).

SYRACUSE, New York
Hancock Intl Airport, 6.5 mi NE

$10.50. 10-15 min.

Ajax, American Intl, Avis, Budget, Dollar, Hertz, National, Snappy, Thrifty.

TACOMA, Washington See SEATTLE/TACOMA, Washington.

TAIPEI, Taiwan

Chiang Kai-Shek Intl Airport, 25 mi (40 km) S

NT$1000 ($38). 40-50 min.

Taiwan Motor Transport Co. every 15-20 min. Pay NT$72 ($2.80) fare before boarding. Two routes: one to Miramar Hotel, Taipei Airport; other to Ambassador Hotel, Taipei train station, Chung Lueu bus station. First bus 7:05 am, last 11:30 pm. 45-60 min ride.

Avis & 12 others.

P NT$240 ($9.15)/day.

TALLAHASSEE, Florida

Tallahassee Regional Airport, 8 mi SW

$10.60, addl riders 50¢ each. 15-20 min.

Avis, Budget, Dollar, Hertz, National.

TAMPA-ST. PETERSBURG, Florida

Tampa Intl Airport, 5 mi NW of Tampa; 15 mi NE of St. Petersburg

To Tampa $13-15, addl riders 25¢ each, 10-15 min. **To St. Petersburg** $34 flat rate downtown, 30-35 min. $44 flat rate to Beach.

To Tampa Central Florida Limousine, $11. Service at all times.

To St. Petersburg The Limo, $11.25. Meets all flights.

To Tampa No. 30 Downtown bus from red departure level every 35-40 min 6:08 am-8 pm M-F; hourly 6:45 am-7:35 pm Sat, 8:58 am-6:04 pm Sun. 85¢. 30-min ride downtown. **To St. Pete/Clearwater Airport** The Limo, $11.25. Meets all incoming flights. Info: 813-572-1111.

Alamo, American Intl, Avis, Budget, Dollar, General, Hertz, National, Payless, Thrifty, and others.

P ST $12, LT $7.

TAMPICO, Mexico

Gen. F. Javier Mina Airport, 5 mi (8 km) NW

$3, 15-20 min.

Avis, Hertz.

TEESSIDE, England

Teesside Intl Airport, 6 mi (10 km) E of Darlington, 13 mi (21 km) W of Middlesborough

£6 ($10.50) to Darlington, 12 min; £12 ($20.90) to Middlesborough, 25 min.

Local diesel every 3-4 hrs. To Darlington, 14 min; Middlesborough, 20 min.

To Darlington £1.05 ($1.80), 20 min. Operates every 30 min.

Avis, Hertz, Godfrey Davis/Europcar.

TEGUCIGALPA, Honduras

Toncontin Airport, 5 mi (8 km) SE

To Prado, Plaza, Honduras Maya hotels, £15 ($5.40), 15-20 min.

Blitz, Budget, Molinari.

TEL AVIV, Israel

Ben Gurion Intl Airport, 12.5 mi (20 km) E

Sh36 ($15.85) to Tel Aviv. After 9 pm 25% addl. Baggage Sh1 (45¢). Tip 10-15%. Fares set by Ministry of Tourism—look for posted rates. It's easier to pay in US dollars than Israeli shekels.

United Tours bus every 45 min 4 am-midnight. $2, 20-30 min. Serves Dan, Sheraton, Ramada, Diplomat, Hilton, Plaza, Carlton hotels, RR station. A/C.

Avis, Budget, Eldan, Hertz.

TENERIFE, Canary Islands

Reina-Sofia Airport, 37 mi (60 km) SW of Santa Cruz de Tenerife

Taxis available for hire to any part of island. Trip time to Santa Cruz 40 min. Determine fare in advance.

Avis, Cicar, Intercanarias, Reisen.

TERRE HAUTE, Indiana

Hulman Regional Airport, 6 mi E

$8.50 to downtown, Indiana State University.

Avis, Hertz, National.

P Free.

TIJUANA, Mexico

Gen. Abelardo Rodriguez Airport, 10 mi (16 km) NE

$7, 15-20 min.

Alliance, Avis, Budget, Hertz, National.

TOKYO, Japan

Haneda Airport, 18 mi (29 km) S

Y5400 ($35), tip not required. 35-40 min to city center.

Monorail, Y300 ($1.80). Service every 5 min 5:20 am-11 pm. 15 min to Hamamatsu-Cho. Taxi from city station.

Budget, Toyota.

TOKYO, Japan

New Tokyo Intl (Narita) Airport, 42 mi (67 km) E

Y20,000 ($130) including tolls, tip not required. 60-90 min downtown. The Japan National Tourist Organization advises: "Taxis are only for the well-to-do...trunk space is small, and the front seat in most cases has to accommodate personal luggage or bags when there are two or more adult passengers."

Airport Limousine Bus every 5 min peak times, up to 20 min other times. Purchase ticket, Y2700 ($17.55), before boarding. Service from 6:50 am-midnight. Travel time 80 min to Tokyo City Air Terminal at Hakozaki-Cho. From there it's a cab ride to hotel or business.

Airport Limousine Bus or Shuttle to downtown hotels. Fares range Y2600-2700 ($17-17.50). Buy ticket before boarding. Departures at least hourly. Travel time 70 min in normal traffic, longer in rush periods. Direct service to hotels in Ikebukuro, Shinjuku, Akasaka, Ginza, Shiba, Shinagawa, Haneda areas.

6-min shuttle bus to Keisei Line Narita Station. Shuttle Y190 ($1.25). Choice from Keisei Station: Skyliner Express, 60 min, Y1750 ($11.40); or Limited Express, 75 min, Y1030 ($6.70). Both terminate Keisei Ueno Station. Purchase tickets at airport counter. Train service every 30 min. Comfortable, baggage space. Plenty of cabs available in city.

To Japan Railway (JR) Tokyo Station JR Shuttle from airport to JR Narita Stn. Y370 ($2.40), 25 min. Choice of trains at station: JR Rapid, 80 min, Y1630 ($10.60); JR Limited Express "Ayame," 60 min, Y2890 ($18.80). **To**

Haneda Airport Limo every 30 min. Y2700 ($17.50). 1 hr 40 min. **To Yokohama City Air Terminal** Limo every 30-60 min. Y3100 ($20.15) 2 hrs.

Avis, Nippon, Toyota.

TOLEDO, Ohio

Toledo Express Airport, 17 mi SW

$28, 20 min by expressway.

Airport van, $15 first passenger, $5 each addl. 30-min run. AMEX, MC, VISA.

Avis, Budget, Hertz, National, Snappy.

P ST $8, LT $4.

TOPEKA, Kansas

Forbes Field Airport, 7 mi S

$10.80, 15 min.

Alamo, Avis, Budget, Enterprise, Hertz, National.

TORONTO, Ontario

Lester B. Pearson Intl Airport, 20 mi (32 km) NW

C$33 ($29), 30-45 min. Tip 10-15%. **Questions?** Look for Ground Transportation/Transport de Surface desk, lower level.

Airport Express Gray Coach every 20 min 6:45 am-12:45 am. Buy ticket in advance at airport or hotel. Fare C$10 ($8.80) OW, C$17.25 ($15.15) RT. 20-35 min to Harbour Castle Westin, Royal York, L'Hotel (Convention Centre), Sheraton Centre (Hilton Intl), Holiday Inn (Bus Terminal), Delta Chelsea Inn.

Airport Express Gray Coach to Islington subway station. Departures every 30-40 min 7 am-12:40 am. Buy coach ticket in advance, C$4.50 ($3.80) OW, C$7.50 ($6.40) RT. Subway fare C$1.30 ($1.15). Toronto subway is clean, easy to navigate with luggage. 20-min ride to Yonge & Bloor, downtown.

Avis, Budget, Hertz, HOJ, Thrifty, Tilden.

P ST C$13 ($11.45), LT C$8.50 ($7.50).

To Woodstock, London, Strathroy, Sarnia Robert Q's Airbus at half-hr intervals 8 am-12:30 am. Info, res: 519-673-6804. **To Woodstock, London** Aboutown Transit London every 2-3 hours 8:30 am-12:30 am. 519-663-2222.

TORTOLA, British Virgin Islands

Beef Island Airport, 8 mi (13 km) E

$4 per person, 30 min. Tip $1.

Anytime, Avis, Budget, Speedy's.

TRAVERSE CITY, Michigan

Cherry Capital Airport, 3 mi SE

$4.50, 10 min.

Avis, Budget, Hertz, National, Thrifty.

P $2/day first 5 days, $1/day thereafter.

TRINIDAD/TOBAGO

Piarco Intl Airport, 16 mi (25 km) SE of Port of Spain

TT$65 ($18.20), 35 min to Port of Spain. To San Fernando, TT$110 ($31), 45 min; to Arima, TT$40 ($11.20), 15 min. Tip discretionary but TT$5 ($1.50) minimum acceptable.

Public Transit Service Corp bus from front of terminal every 30 min 5 am-11 pm. TT$1.50 (40¢), 30 min to Port of Spain. No baggage.

Auto Rentals Limited, Himraj Taxi & Rental, Singhs.

P TT$25 ($7)/day.

TUCSON, Arizona

Tucson Intl Airport, 12 mi SE

$12-14. 15-20 min ride.

$10.50 to downtown hotel. To resorts, $17. To private residences, $12.50 first person, $7-9 each addl same address. Info & res: 602-889-9681.

Ajax, Alamo, American Intl, Avis, Budget, Dollar, Enterprise, Hertz, National, Snappy, Thrifty.

P ST $7, LT $4.40/day. Remote lot $3.

TULSA, Oklahoma

Tulsa Intl Airport, 9 mi NE

$12.75, addl passengers 75¢ each. 15-20 min.

To Bartlesville Phillips Petroleum Co. operates a free limousine for employees which is also available to other travellers as space permits. Hourly at :15,

9:15 am-10:15 pm, M-F. Sun at 11 am, 12:30, 2, 3:30, 5, 6:30, 8 pm. No Sat service. Check in at Phillips Lounge, arrivals area. Info: 918-661-1533.

Avis, Budget, Dollar, Hertz, National.

P ST $12, LT $5/$2.

VAIL, Colorado
Vail/Eagle County (Vail/Beaver Creek) Jet Center, 35 mi (56 km) W

$14/person or $82, 1-6 passengers, 35 min to Vail. Vail Valley Taxi: 303-476-8294.

Vail Valley Transportation: 800-882-8872. (Service available to Denver Stapleton.)

Budget, Hertz, Thrifty.

VALLETTA, Malta
Luqa Airport, 4 mi (6 km) SW

Always available, 15 min. £5 ($1.60).

No. 32, 34, 35 bus every 20 min 6 am-10 pm. 10¢.

Avis, Hertz.

P Free.

VANCOUVER, British Columbia
Vancouver Intl Airport, 11 mi SW

C$21 ($18.50), 23-30 min.

Perimeter Transportation, C$8.25 ($7.25) to downtown hotels. 20-30 min. Airport Limousine Service, C$26 ($22.90).

ABC, Avis, Budget, Hertz, Thrifty, Tilden.

P ST C$8.50 ($7.50), LT C$3 ($2.65).

VENICE, Italy
Marco Polo Airport, 8 mi (13 km) NE

Lit16,000 ($12.80), 15 min to Piazzale Roma, where one transfers to a vaporetto mini-ferry or a water taxi. Tip landside driver 5-10%. **Water taxi** Lit90,000 ($72) for door-to-door service. One fare covers up to 8 people.

Water bus Lit13000 ($10.40) per person. Set route includes Rialto, Piazza San Marco, some major hotels. About 45 min.

ATVO bus coincides with flight schedule. Buy ticket for Lit2500 ($2) at ATVO counter in airport. Plenty of baggage space. 20 min to Piazzale Roma.

Autorent, Avis, Europcar, Hertz, InterRent, Italy by Car/Budget, Maggiore.

VERACRUZ, Mexico

Gen. Heriberto Jara Airport, 5 mi (8 km) SW

$4, 15 min.

5-passenger colectivos available for less than cab.

Avis, Budget, Dollar, Hertz.

VIENNA, Austria

Schwechat Airport, 10 mi (16 km) SE

S320 approx. ($27), 20-30 min. Bags S10-20 (85¢-$1.70) each depending on weight.

Vienna Airport Service 24 hour service every 20 min to midnight, 30-60 min after. Baggage carried. 20-30 min to City Air Terminal (Hilton). Pay S50 fare ($4.20) when boarding. Return buses marked "Flughafen."

Hourly service 3 am-10:40 pm, to Wien Sudbahnhof (south RR station) and Wien Westbahnhof (west station); buses so marked. Baggage racks. Board directly outside terminal. Pay S50 ($4.20) to driver. 20-35 min trip.

First train 5:11 am M-F, 5:34 am Sat, 6:15 am Sun. Last train 10:05 pm. Frequent morning service then hourly from 8:05 am. Comfortable, good baggage racks. 30-min trip. Buy ticket for S30 ($2.40) before boarding.

Avis, Budget-Leihwagen Union, Denzel, Europcar, Hertz, Intercity, InterRent.

P ST S117 ($9.25), LT S72 ($5.70).

WACO, Texas

Waco Municipal Airport, 7 mi NW

Cabs do not wait at airport, must be called. 12 min downtown, 15 min to Baylor area. Fares $7-8.50.

Avis, Hertz, National.

WARSAW, Poland

Okecie Airport, 6 mi (10 km) SW

Zl70-80,000, 30 min. Tip 10%. Taxi stand is reportedly opposite the arrivals building, but travellers report taxis sometimes hard to find. Phone number in Warsaw for radio taxi: 909

Every 30 min 5 am-11 pm to LOT office at Warynskiego 9. Fare Zl2000.

No. 175 every 10 min 5:30 am-10:50 pm. Zl2000. Overnight service on No. 611 every 30 min 10:52 pm-4:25 am. Buy tickets at RUCH kiosk in departures hall, across street from arrivals hall.

Orbis Rent a Car, in arrivals hall and Forum Hotel, handles bookings for Avis, Europcar, Hertz, National.

WASHINGTON, D.C. (Dulles)

Dulles Intl Airport, 26 mi W of Washington, DC

$38-40 to midtown DC, 45-60 min. National Airport $40. For fares to other places, call dispatch: 703-471-5555.

Washington Flyer to downtown D.C. terminal at 16th & K St. and principal hotels, 45 min. Every 30 min 6 am-midnight, $14. Info: 703-685-1400.

Washington Flyer to West Falls Church Metro Station every 45 min 6 am-10 pm M-F, Sat from 8:15 am; Sun from 10:15 am. Last Sun departure at 5 pm. 17-min ride. Transfer to Orange Line Metro. $1.65, 20-25 min to DC downtown stations.

To suburban Maryland Washington Flyer service to Bethesda, Gaithersburg, Rockville, and various hotels leaves hourly on the half hour 5:30 am-9:30 pm. $15-$20.

To National Airport Washington Flyer hourly 5 am-11 pm, $12. 45 min.
To BWI Take Washington Flyer to 16th & K terminal, transfer to BWI van. See Baltimore listing.

Alamo, Avis, Budget, Dollar, Hertz, National, Thrifty.

P ST $24, LT $7, Satellite $5.

WASHINGTON, D.C.

Washington National Airport, 4 mi S

$9-14. Cabs licensed in D.C. do not have meters. Fare is based on distance traveled by zones. To avoid surprises, confirm in advance with driver. Virginia, Maryland cabs metered: Pentagon, $7.50; Rosslyn, Alexandria $9; McLean, $23; Tysons, $18; Dulles Airport, $38; BWI, $45; Baltimore downtown, $50.

Washington Flyer to suburban Maryland, Bethesda, Gaithersburg, Rockville and various hotels. Fares $14. Hourly on the half hour 7:30 am-10:30 pm. Info: 703-685-1400. Washington Flyer to downtown Washington and various hotels. Fares: $7.

Metro 6 am-midnight M-F; 8 am-midnight Sat; 10 am-midnight Sun. Station opposite N Terminal. There's a shuttle bus but it's faster to walk: 5-6 min. Buy ticket from machine. Rush-hour fares $1.20, (non rush-hour $1) on Blue or Yellow Line. Trains every 5-10 min. Allow 40 min total from airport to midtown destinations. Clean, quiet, comfortable coaches. Good choice if you're traveling light.

No. 13 fill-in service to Pentagon, DC, on Sat & Sun mornings before Metro runs. Sat every 15 min 6:14-7:44 am; Sun every 30 min 6:22-9:48 am. Picks up at airport Metro station. $1.70 exact change. 11 min to Pentagon, 24 min to 10th St. & Pennsylvania Ave. N.W. **To Mt. Vernon, Ft. Belvoir** No. 11 bus from Metro station every 30-60 min 6:17 am-7:31 pm; 7:37 am-7:07 pm Sat; 9:29 am-6:35 pm Sun. $2.05 exact change rush hour, $1 nonrush.

To Dulles Intl Airport Washington Flyer hourly 6 am-10 pm. $14. 35-40 min. **To BWI, Baltimore** Take cab to 16th & K St. bus terminal in D.C. Transfer to BWI bus. Hourly service, $13. Info: 301-859-7111. 65 min to BWI.

Alamo, Avis, Budget, Hertz, National.

P ST $20, LT $8.50/$7.

To Fredericksburg Groome Transportation, 800-552-7911 (Va). **To Virginia locations** Springfield, Ft. Belvoir, Quantico, FBI Academy, Stafford, Manassas: DAFRE Limo, 703-690-3102.

WELLINGTON, New Zealand

Wellington Intl Airport, 4 mi (6 km) SE

NZ$12 ($7), 20 min. To Burma Lodge, NZ$20 ($11.70), 25-30 min.

Vickers Coach Lines every 20 min 6:30 am-9:50 pm M-F. NZ$4.50 ($2.65) paid before boarding. Sat 7:30 am-8 pm, Sun 7:30 am-10 pm, at 30-min intervals. Stops at any bus stop on request, RR station. Downtown, 20 min.

Supershuttle on demand. Up to 10 passengers in one bus. NZ$8 ($4.65) reduces with number of passengers carried.

Avis, Budget, Hertz.

P NZ$12 ($7)/day.

WEST PALM BEACH, Florida
Palm Beach Intl Airport, 5 mi W of The Breakers

$7-8, 10-15 min.

Avis, Budget, Hertz, National.

P ST $8, LT $3.50.

WHITE PLAINS, New York
Westchester County Airport, 4 mi NE

Downtown White Plains $15, 10-min ride. If no cab at airport call 592-8534 or 949-0110. Other fares: American Can, MCI $12; Armonk $16; Arrowwood $12; Chappaqua, Elmsford $21-27; Greenwich $23-25; JFK $55 plus tolls; LaGuardia $50 plus tolls; Pepsico $10; Rye Hilton $15; Stamford $25-32; Stouffer $14; Tarrytown $22.

Connecticut Limousine service to White Plains, New Rochelle, Rye, Tarrytown, Elmsford, Riverdale-Bronx. Info: 914-699-1000.

To New York City Metro North Commuter train to Grand Central Terminal $6 peak, $4.50 off-peak. Frequencies from 10 min rush hours to 60 min late night, weekends. Operates 5 am-midnight. Info: 212-532-4900.

Avis, Budget, Dollar, Hertz, National, Snappy.

P $4.75/day.

WICHITA, Kansas
Wichita Mid-Continent Airport, 6 mi SW

$7.50, addl passengers $1 each. 10-15 min.

Avis, Budget, Dollar, Enterprise, Hertz, National, Thrifty.

WILLEMSTAD, Curacao
Curacao Intl Airport, 4.5 mi (7 km) NW

Naf18 ($10), 18 min; to Princess Beach Hotel, Naf19 ($10.50), 25 min; to Caribbean Hotel, Naf16 ($9), 15 min. Taxis have TX on plate and Taxi sign on roof. One fare covers 4 passengers; 5 adds 25% to fare, 6 adds 50% to fare. 25% surcharge 11 pm-6 am. Baggage: if unable to close trunk, $1 each extra bag.

ABC Co. bus departs hourly 6 am-11 pm next to arrivals hall. Pay driver Naf70 (40¢). Baggage OK. To Otrabanda, 20 min; to Punda, 1 hr.

Avis, Budget, Caribe, Hertz, National, Ric-Car, Ruiz, Uralco.

P Naf8 ($4.50)/day.

WILMINGTON, Delaware
New Castle County Airport, 5 mi (8 km) S

$14, 10-12 min. If no cab at airport, use courtesy phone.

No. 22, $1.15. Frequent, 24 hours a day.

Avis, Budget, National.

WINNIPEG, Manitoba
Winnipeg Intl Airport, 5 mi (8 km) W

C$11 ($9.70), 15 min.

Ambassador, Executive, Kidd, Leisure London, C$9 ($7.90) to downtown hotels, 20 min.

No. 15 Sargent bus about every 20 min. C$1.15 ($1). 20-min ride downtown.

Alpine, Avis, Budget, Hertz.

YOUNGSTOWN, Ohio
Youngstown Municipal Airport, 14 mi N

$18 flat fare, 30 min to downtown Youngstown.

Avis, Budget, Hertz, National, Snappy.

ZAGREB, Yugoslavia
Zagreb Airport, 10.6 mi (17 km) SE

$25, 20 min.

Eurcont bus every 30 min, 6 am to last flight. 25 min to JAT air terminal, Central Bus Terminal. $3. Hotel Zagreb Inter-Continental operates courtesy transport to and from airport.

No. 267 to ZET terminal (main RR) every 1-2 hrs early morning, afternoon-evening. $1.

Avis, Budget, Dollar, Hertz, Unis.

ZURICH, Switzerland
Zurich Airport (Kloten), 7.5 mi (12 km) N

SF40 ($27.60), 15-20 min depending on traffic. If fare includes tip, a sign near meter will say so. Otherwise tip 10%.

Look for railway sign outside customs—station is beneath airport. Train every 10-20 min 6:06 am-midnight. 10 minutes into Zurich. Excellent service. Fares SF4.20 ($2.85) 2nd class. **Direct rail services** from airport station to many destinations in Switzerland including Lucerne, Lausanne, Geneva, Bern, Interlaken, Chur. **Fly Rail Baggage** service to and from 100+ rail stations all over Switzerland operates at Zurich, Geneva & Basel Flughafen stations. Time limits apply and there is a charge of SF10 ($6.90) per bag.

No. 768 bus to city every 9 min rush hours, 15-30 min other times. SF3.60 ($2.50).

Avis, Budget, Europcar, Hertz, InterRent.

P ST SF16 ($11), LT SF8 ($5.50).

NOTES